# GREAT FEUDS IN TECHNOLOGY

*Ten of the Liveliest Disputes Ever*

Hal Hellman

John Wiley & Sons, Inc.

Published by John Wiley & Sons, Inc., Hoboken, New Jersey
Published simultaneously in Canada



For general information about our other products and services, please contact our Customer Care Department within the United States at (800) 762-2974, outside the United States at (317) 572-3993 or fax (317) 572-4002.

Wiley also publishes its books in a variety of electronic formats. Some content that appears in print may not be available in electronic books. For more information about Wiley products, visit our web site at www.wiley.com.

ISBN 0-471-20867-1

Printed in the United States of America
10 9 8 7 6 5 4 3 2 1

*To Jill, Jennifer, Stephanie, and Scott—*
*family (loved), friends (useful), and neighbors (delightful).*

# Contents

# ACKNOWLEDGMENTS

A book like this requires sorting through an enormous amount of material. The Internet was truly an amazing, if sometimes frustrating, resource. Yet, for greater depth, printed books and libraries still proved essential for much of my research. Most helpful have been the collections of historic materials at the New York Public Library and the newer Science, Industry, and Business Library, both in Manhattan; the Dibner Library in Cambridge, Massachusetts; and the Archives of the Smithsonian Institution in Washington, D.C.

I would also like to thank the staff at my own local library in Leonia, New Jersey, which is, happily, part of a countywide library system. Special thanks to Teresa Wyman, reference librarian, with whose help I was able to retrieve useful materials from across the country. Thanks also to Annemarie Mascarenhas, reference librarian at Bergen Community College, who went out of her way to be helpful.

Visits to several science museums and special collections provided some useful background. These included the Old Rhinebeck Aerodrome in New York State; the Edison National Historic Site in West Orange, New Jersey; the American Museum of Natural History in New York City; and the Liberty Science Museum in Jersey City, New Jersey.

Many colleagues and friends have been helpful. Some of this had to do with a willingness to listen to my gripes and groans during periods of frustration and catastrophes, both major and minor. Equally important was their help in answering questions and/or reading and commenting on parts of the manuscript as it was being generated. Among those in the latter category:

Walter Licht, Department of History, University of Pennsylvania; Manfred Stutzer, collector of memorabilia for miners' safety lamps, Ludwigshafen, Germany; David M. Knight, Philosophy Department, University of Durham, U.K.; Marc Rothenberg, editor, Joseph Henry Papers Project, Smithsonian Institution Archives; David Hochfelder, assistant editor of the Thomas Edison Papers at Rutgers University; Doug Tarr, research archivist, Edison National Historic Site; Bob Arnebeck, automobile historian; Rosalynn Driskell, Ford Motor Company Public Affairs, George Skoler, retired patent attorney; William Cox, associate archivist, Smithsonian Institution Archives; Alexander B. Magoun, director, David Sarnoff Library; Ray Ettington, docent at the Glenn H. Curtiss Museum of Early Aviation in Hammondsport, New York; Jack Carpenter, aircraft historian; Donald G. Godfrey, broadcast historian and professor, Arizona State University; Albert Abramson, television historian; Frank Duncan, naval historian; Steve Rozen, research scientist, Whitehead Institute, Massachusetts Institute of Technology; Larry Thompson, chief, Communications and Public Liaison Branch, National Human Genome Research Institute; Mark Guyer, director of the Division of Extramural Research at the National Human Genome Research Institute; Bruce M. Chassy, assistant dean, Office of Research, College of Agricultural, Consumer and Environmental Sciences at the University of Illinois, Urbana.

Additional and special thanks go to my editor, Jeff Golick, who helped me through some difficult times; to my agent, Faith Hamlin, for her support; and to my daughters, Jill and Jennifer, who lent emotional support when I needed it most. Jennifer also provided some very useful editorial comments on a number of the chapters.

# INTRODUCTION

A thousand years ago, the mighty King Canute ruled over England, Denmark, Norway, and Scotland. King Canute was considered, if not quite a god, close to one. He not only wielded absolute power over his subjects, he was often thought of as a ruler of nature, as well.

His subjects, abject and superstitious, thought there was nothing he could not do. "Oh, King, is there any one as mighty as you?" "There will never be a being as great as you." And so on. Over and over. Did he eat it up? Or did he get a little tired of it?

Two legends have come down to us from that time. Both start off with Canute and his court down at the seashore. Both begin with his throne being placed before the incoming tide. In the first legend, he is so smitten by all the flattery and adulation that he thinks he is actually as great and powerful as all his subjects say he is. He feels he can even control nature and has gone down to the seashore to demonstrate his powers over the incoming tide. He raises his hand and says, "Sea, I command you to retreat!"

Of course, the sea pays no heed, and old Canute has to beat a hasty retreat. This legend makes him out to be an arrogant fool, one so besotted with his own powers that he oversteps all sensible bounds and is slapped down by nature.

But Canute appears to have been a sensible and effective ruler, and from that point of view the legend really doesn't make sense. Though flattered by all the admiration and subservience, he can see a negative side to it, as well. If harvests begin to weaken, for example, he fears that his subjects, instead of redoubling their efforts, might say, "No sweat. The king will handle it." He recognizes this as a potentially serious problem.

And so, to demonstrate that he does *not* have supernatural powers, in the second legend he takes his court down to the seaside, and when the tide does not respond, he says, "You see. I *cannot* control nature."

It's virtually the same story as the first legend, but with quite a different meaning and implication. It would appear that the first interpretation, the more common one, is basically an inversion of history. What could have caused it? In the search for an answer, let's compare the story with another that takes place eight hundred years later.

At the beginning of the nineteenth century, the textile hand workers in England are seriously troubled by the introduction of new machinery. Proud of their skills and their handiwork, a large group gathers and goes around breaking the offensive new machines whenever the factory owners try to introduce them. For reasons that will become clear in chapter 1, they have come to be called Luddites. Over time, the term has become synonymous with machine-breakers or, more generally, technophobes—people who fear or hate new technology.[1]

There are such people today, and they wear the label *Luddite* proudly. More commonly, and among those who feel more positively about modern life and modern technology, the same term is applied to such people with disdain. For example, Thomas R. DeGregori, professor of economics at the University of Houston, refers to a group "whose brains were eaten away by a Luddite ideology."[2]

DeGregori is using the word *Luddite* in its commonly accepted sense. What's interesting, though, is that, as in the Canute story, this interpretation is a misreading of history. For, although the Luddites were indeed machine-breakers, we shall see that the underlying drive had more to do with economics than with a simple fear or hatred of new technology.

Both stories have been subverted for similar reasons.

In Canute's case, the change came about because people throughout history have wanted gurus, have wanted magic, have wanted someone, or something, to do their thinking for them and to take over when problems arise. But legends, like commodities and new technologies, only last if they satisfy the consumer.

Canute tried to show he was not all-powerful, that he could not control nature. But following generations didn't want to hear that. While many people want gurus, they also like to see the mighty topple. So the tale has become one in which Canute overreaches, tries to exert mastery over nature, and gets knocked down. As a result, of the two legends, the second is more common.

This idea—that nature is good and human attempts at mastering it are bad—has appeared over and over again throughout history. Goethe's *Faust* and Mary Shelley's *Frankenstein* are just two of the better-known examples. In the 1920s, the Agrarian literary movement of the American South decried the rise of industrialism.

The current incarnation is the modern Luddite. But using the original Luddites as a model does them a serious injustice. Granted, they were responding violently to a major challenge to their traditional ways. But the real issues were economic, with poverty and starvation very real threats and motivations.

In the case of the Luddites, there are not two legends, but two sides to the same legend. Similarly, there were from the beginning those who supported the machine-breakers, and those who felt they were the scum of the earth. As the Industrial Revolution proceeded, and especially in our own day, supporters of new technologies needed a convenient label for those who fought these technologies. The Luddites, unwittingly, supplied it. And, as in the early nineteenth century, modern Luddites have taken on the label for themselves, as well—again doing the original Luddites an injustice.

There is, of course, a wide range of thought and opinion among the pure technophiles and the pure neo-Luddites. And there are certainly serious issues that must be carefully evaluated when a truly new technology is on the verge of being introduced.

In this book we look at a variety of new technologies and how they were greeted over the past couple of hundred years. Perhaps the variety of responses described here can help us better compare the benefits and risks of a new technology and give some insight into how society might better deal with it.

Each chapter also describes a different sort of battle that erupted as a result of such an introduction, for, as in both science and medicine, when a new technology is introduced, it is likely to be stepping on the toes of an earlier one. At the very least, questions of priority may provoke some acrimonious disputes, as in the case of Davy versus Stephenson, where Davy's supporters can't believe that a self-educated engineer like Stephenson could have come up with a workable, and desperately needed, safety lamp for miners (chapter 2).

Both Davy and Stephenson ask for nothing more than the credit for their invention. That is unusual. More often, the invention of a new technology brings on bitter commercial battles. In chapter 3, for example, Samuel Morse works long and hard to bring his telegraph into existence. But he uses others' ideas and gets help from outside. When the commercial value of the telegraph finally becomes apparent, challenges to his invention come from every side.

In addition, however, one of the challenges—from the highly respected American scientist Joseph Henry—concerns a matter of pride rather than economics and plays a part in swinging American technological creativity from pure tinkering to science-based development.

Thomas Edison (chapter 4) also changes the course of technology, and in several ways. It is he who shows that electricity can be put to work in the service of humanity. Yet, in his battle with George Westinghouse over what kind of electricity the country would run on, he chooses the wrong side. Nevertheless, he transforms the domain of invention from that of the individual tinkerer to today's organized industrial research organizations.

In chapter 5, Henry Ford starts off as an underdog who, in his tenacious struggle to mass produce automobiles, must battle a patent lawyer, an East Coast conglomerate, and a patent that gets changed more times than a baby's diaper.

By chapter 6, we take to the air, but not until the Wright brothers have found a way to control a heavier-than-air craft. Others, including Glenn H. Curtiss, maintain that the Wrights should not get the credit and do everything in their power to break the Wrights' patent.

Another major patent battle erupts in chapter 7 over the question of who should get the credit for, and who should profit from, the creation of television. Should it be Philo T. Farnsworth, the one who invents it, or David Sarnoff, who turns it into the major technology it has become in our day?

In chapter 8, we change course and meet Hyman Rickover, who takes on the entire Navy, including chief of naval operations Elmo Zumwalt, in his attempt to create a battle fleet powered by nuclear propulsion.

The human genome is the star of chapter 9. Two groups, one led by Francis Collins and the other by an upstart challenger, Craig Venter, get mixed up in a priority dispute of major proportions.

In the final chapter, we come full circle; we deal with the question of whether Jeremy Rifkin's ongoing crusade against genetically modified foods is an example of neo-Luddism or a valid campaign against a technology that threatens our very existence.

Some readers will wonder why a book on technology disputes does not include a chapter on computers. I certainly considered including material on Bill Gates versus Steve Jobs of Apple Computers or Larry Ellison at Sun Microsystems. The problem was that those stories seemed more concerned with business decisions, indeed with questions of monopolistic practices, and therefore lacked the qualities of a good feud. Other disputes just seemed more dramatic, or more concerned with the technologies per se, or both, and therefore more interesting to me.

We begin, then, with the story of the Luddites. It concerns a struggle whose effects still resonate today and helps set the stage for the controversies that follow.

# CHAPTER 1

# *Ned Ludd versus the Industrial Revolution*

## Are Machines the Problem?

Sherwood Forest. Who does not associate Sherwood Forest with Robin Hood—that legendary, daring hero of medieval England?

Though he was always on the wrong side of the law, he invariably is seen as heroic; for although he and his "Merry Band" stole from the rich, it was, says the thirteenth-century legend, so that he could give to the poor. Members of his band, including Little John, Friar Tuck, and the gentle, lovely Maid Marian, also achieved fame.

Seven hundred years later, another legendary character, Ned Ludd, emerged from Sherwood Forest. There are surprising similarities between Robin Hood's Merry Band and Ned Ludd's group, collectively called Luddites, but also some important differences. There is, for example, no love interest. More important, the Luddites were no Merry Band. Theirs was serious business. In fact, they were all about business, though from a decidedly negative point of view.

### *A Volcanic Eruption*

The Luddite saga begins March 11, 1811, in Nottingham, a small town that was not far from Robin Hood's base of operations in Sherwood Forest. It was now, however, an important center for the production of cotton hosiery and lace, and throughout that winter day, framework knitters streamed in from their homes and workshops in the surrounding countryside, where they had been operating their hand looms for years.

Ironically, speeded-up and power-driven equipment, including power looms, wide stocking frames, and knitting machines, had been entering the textile industry for several decades, mostly without incident. But recent trade problems had led the owners to decrease payments for the work, and this, combined with the rising cost of food, was driving the workers into poverty and starvation. All of this had brought matters to a head at Nottingham, and the disgruntled workers were complaining to the sympathetic townspeople; but they directed their complaints mainly against the owners they worked for, who both underpaid them and called for "cut-up" stockings. These were knitted in wide sheets, then cut up and sewn into stocking form. Inferior in quality to true knitted stockings, they not only were cheaper but could be made using unskilled labor and the (more expensive) type of stocking loom called a wide machine.

Feelings ran high, and the men demanded that something be done. A variety of "authorities," mainly dragoons from the Crown but also hired hands paid by the owners, ranged the streets and tried to maintain order. At about 9 P.M., the crowd finally dispersed, at which point the townspeople and the owners heaved a sigh of relief.

But the real trouble was about to begin. The unhappy workers, taking a leaf out of Robin Hood's book, had simply disappeared into the darkness. A particularly unhappy group marched to nearby Arnold, and in the dead of night, proceeded to break into homes containing the hated machines that had been rented from the offending owners. By daylight, some sixty large stocking frames had been destroyed.

The explosive release of anger continued over the following weeks. It was strike by surprise, then disappear into the night. The owners, so spread out that they had no way to "circle the wagons," found it difficult—in most cases, impossible—to defend against the raids, which could occur almost anywhere and anytime, sometimes even during the day. Often the raiders attacked in several different areas in the same night. It's important to note that, at least in the early stages, they targeted specific owners, and specific machines.

Throughout, the aggrieved workers knew they had to explain their actions. Aside from verbal complaints to anyone who would listen, these explanations came mainly in the form of written notices to offending parties and/or proclamations aimed at the public and at the Crown. All were signed by "Ned Ludd," or sometimes "General Ludd," or even "King Ludd." The group apparently took the name from a story often told in the area about a young man called Ned Ludlam. There were many different versions.

In one version, he was an apprentice knitter and perhaps of weak intellect. One day he was ordered by his father to get on with his work. Responding in a fury, he grabbed a hammer and smashed his frame into pieces.[1] The *Shorter Oxford English Dictionary* (1964) describes him as insane. In a different version of the story in the 1902 edition of the *Encyclopedia Britannica,* he was the butt of boys' pranks in the village, and on one fateful day he pursued one of his tormenters into a home that housed two of the frames used in manufacturing stockings. Not able to catch the boy, he took out his anger on the frames.[2]

Though, as with Robin Hood, there is little hard evidence of a real person by the name Ned Ludd, no one doubts that his followers were real. Over the two years following the opening foray in March 1811, the various Luddite raiding groups—apparently well organized and disciplined—caused widespread damage to machines and property, amounting to over 100,000 pounds, an enormous sum in that time.[3]

As 1811 wore on, trade continued to worsen, and farmers experienced a poor harvest, leading to still higher food costs. Though some owners did raise their payments, it was nowhere near enough. By November of that year, the raids had become even more virulent, and the raiders began to take on the mien of an organized band. Led by someone—it has never become clear who—they set off on November 10 for Bulwell and the factory of Edward Hollingsworth, an owner particularly hated by the knitters. Though Hollingsworth had expected trouble and had tried to fortify his factory, the ferocity, organization, and effectiveness of the raid caught him by surprise. In the confusion, there were shots, and a Luddite, John Westley, was killed. Nevertheless, the group overran the defenders, did their damage, then quietly dispersed and disappeared into the darkness.

It was the first death resulting from the riots, but far from the last.

In the early months of 1812, the raids slackened. E. P. Thompson, in his classic *The Making of the English Working Class,* says the attackers did see some success.[4] Many hosiers agreed to pay higher prices. The government had stationed several thousand troops in the area, however, and a bill to make frame-breaking a capital crime was put forth in Parliament.

## *Escalation*

Even as the rioting quieted down in the Nottinghamshire area, it began to spread to other parts of the textile manufacturing areas to the north, including Yorkshire, a wool center, and Lancashire, which specialized in cotton.

The Yorkshire area saw a particularly bloody confrontation, in word, action, and reaction. William Horsfall, a local textile mill owner, refused to be intimidated and swore he would ride "up to his saddle girths in Luddite blood."[5] Horsfall was busily installing new equipment in his mill at Rawfolds. This consisted mainly of power-driven shears that could easily quadruple the output of the cropper's traditional heavy handheld shears. (Traditional shears—which could weigh up to fifty or sixty pounds—were used to cut off projecting threads and fibers, called the nap, from the finished cloth. The croppers who wielded them were a well-paid and proud group.) Horsfall knew he was a major target, and fortified his mill accordingly—including mounting a cannon in the building, barricading the stairs with spiked rollers, and placing a tub of oil of vitriol (a highly caustic substance) at the top.

Sure enough, a major attack was mounted on April 27, 1812. A force of around 150 men, perhaps more, wielding hatchets, heavy hammers, and guns, attempted to enter by battering down the door, breaking windows, and any other means possible. But Horsfall had done his job well. The attack failed, and eventually the raiders dispersed. But in the attempt, shots were fired by both sides, and two Luddites were killed. The Luddites vowed revenge, and not long after, Horsfall was assassinated while riding his horse along a deserted route.

For several months, despite the widespread employment of government spies along with the presence of some four thousand troops in the area, not one of the Rawfolds attackers was brought up on charges. It was an astonishing show of sympathy by the community. It was also, however, partly a fear of reprisal from the Luddites.

But the government was upping the ante. On December 10, 1812, for example, the county of Leicester issued a "Caution" against all persons engaged in the crime of frame-breaking: "Every person forcibly entering a house in night time, with intent to break a frame, and every person in any manner aiding or assisting others in so doing, is guilty of burglary, which is punishable by DEATH."[6]

Part of the reason for the curious (to our eyes) connection with burglary has to do with the participation of outsiders (non-Luddites) who, looking to cash in on the raiders' rage, engaged in looting. Later riots, particularly in the north, saw even more of this, despite the fact that burglary was indeed often punishable by execution.

Still, smashing machines or other property was one thing. Even deaths during battle could be swallowed. But the assassination of a defenseless man, even one as hated as Horsfall, was another matter. Eventually, and for the first time, information was given that led to the arrest and conviction of some of the perpetrators. This was the first major break, and it led to an additional series of arrests in the three major regions involved. These included some Luddite ringleaders, all of whom were tried before a special commission at York Castle in January 1813. Twenty-four men were judged guilty, and seventeen were executed. Others were transported to Australia.

Unhappily, the situation of those who did give information was often as pitiful as those who were condemned. This included constant fear of reprisals, including ostracism, which in those tightly knit communities was a serious matter.

Though the Luddites continued their attacks, new machines continued to be introduced in the Yorkshire area, and the number of croppers—once an independent, tough, respected group—dropped from more than 1,700 to just a few in five years.[7] The croppers, says Thompson, came closest to the popular image of the Luddite: "They were in direct conflict with machinery which both they and their employers knew perfectly well would displace them."[8]

## *A Difficult Time*

What could have brought the Luddites—generally law-abiding, God-fearing citizens—to such a state of anger? Textile manufacture had played an important role in England's commercial success, in both domestic and world trade. And at the turn of the eighteenth century, most of the manufacturing had still been done by these independent operators. The system had worked well for many years; but by the time of the Luddite activities, a variety of factors had led to a serious decline in the fortunes of the workers. A rapid increase in population in the previous decades had led to more dependence on foreign markets. But wars, including a drawn-out series with France, and increasingly annoyed American reactions to British economic and maritime policies, had led to serious disruptions in the flow of goods. By 1810, the American Congress already included a group called the war hawks who were calling for war against England. By then the United States had become Britain's single largest customer for its products, which made this loss particularly bitter. From October 1810 to March 1811, a million pounds of woolen cloth, manufactured in the north and intended for the American market, had accumulated, unusable and unsalable. All of this had contributed to a serious economic depression. An especially cold winter had added to the workers' problems.

A change in the landholding policies of the country also prevented the workers from carrying out part-time agriculture on pieces of common land, which in the past had helped feed and even clothe them in times of need.

At the same time, the owners were trying out a new kind of production. Leading eventually to what we now know as industrial production methods, it promised economies of scale and efficiency for the owners, but it seemed highly threatening to the workers. Finally, there were, of course, unscrupulous manufacturers who added to the trouble by using a variety of means to take advantage of their position: one method was to defraud the workers, say, in company stores; another was to employ "colts," or unapprenticed workmen, at lower wages; a third method, and one that particularly galled the proud handworkers, was to manufacture inferior goods to bring prices down and increase sales.

All of these factors conspired to make life increasingly difficult for the independent operators. Their situation, in fact, combined the economic risk of the entrepreneur with the powerlessness of the serf; they had the worst of both worlds. Cost of food and raw materials was increasing; need for their labor, decreasing.

Although some of the owner/manufacturers had been willing to at least try to help the knitters, an obstinate core of owners who, used to exercising almost total control, resented the workers' "uppityness." These owners had resolved to hold fast and in some cases even to reduce payments to their workers.

The result was that many of the workers were already in deep financial trouble. In early nineteenth-century England, this meant loss of homes, starvation, and anything else that comes with poverty; for in those days, there were no governmental safety nets such as we know today in Western countries. Malcolm I. Thomis, a British historian who wrote a groundbreaking book on the Luddites, spells out the situation: "The workers had so little cushion that a doubling in price of oats in the northern diet "put hundreds of thousands in a state of desperation. . . . In May 1812, an ironical correspondent suggested in the local press that the present troubles might be cured if doctors would only get together to find out how appetite might be eliminated."[9]

There was also a social factor. As late as 1818, a cotton spinner said, "I know it to be a fact, that the greater part of the master spinners are anxious to keep wages low for the purpose of keeping the spinners indigent and spiritless . . . [as much as] for the purpose of taking the surplus into their own pockets." He added that the textile worker "cannot travel and get work in any town like a shoe-maker, joiner, or taylor; he is confined to the district."[10]

## *Pre-Luddite Years*

The Luddites would have preferred to accomplish their objectives in less violent ways; there had been earlier attempts at negotiating with the owners. One recommendation was a controlled introduction of cropping machinery, with alternative employment, or at least some financial help, for displaced workers.[11] It came to naught. The workers were so spread out that it was hard to create the kind of unions we know today, which might have been able to exert the right kind of pressure. The inevitable result was that the workers as a group had little bargaining pressure, and the onus fell on individual workers.

They faced another barrier: a general resistance to the very idea of their combining forces, which showed up in a series of laws passed in 1799 and 1800. Unthinkable today, these Combination Laws strengthened and consolidated longstanding antiunion leanings by specifically forbidding the organizing or "combining" of workers to achieve higher wages, shorter hours, or better working conditions.

A second important factor is that machine-breaking was not a new idea; there were other well-documented cases before the Luddite era. One took place in 1710, a full century before the Luddite uprising, when workers smashed the machines of a London hose manufacturer who had decided to ignore a guild rule restricting employers to no more than a few apprentices.[12] Many employers then moved to the outlying areas, such as Nottinghamshire, to help them evade regulations such as those put forth by the guilds.

Later, in 1768, a mill owner named Richard Arkwright invented a spinning device that found use in the manufacture of cotton yarn; but in 1788, when manufacturers attempted to apply the machines to spin wool yarn, angry workers not only wrecked the machines but also damaged the building in which they were housed. There also had been cases of break-ins and damage to knitting frames being put to use by owners. In 1779, a worsening period of trade led to riots in Lancashire, resulting not only in destruction of machinery but in burning down a mill completely. Arkwright quickly laid in a great supply of arms at his mills in Cromford and was never directly attacked.[13]

## *Raiders, Rebels, or Victims?*

Although the Luddite movement is related to this tradition of machine-breaking, there are major differences—and different opinions of these differences. Thompson writes that the Luddite movement is distinguished "first, by its high degree of organization, second, by the political context in which it flourished. . . . Luddism," he argues, "was a *quasi-insurrectionary movement*, which continually trembled on the edge of ulterior revolutionary objectives."[14] But the various geographical sections also had somewhat different experiences. Thompson summarizes: Nottingham was the most organized and disciplined; Lancaster experienced the highest political activity; Yorkshire moved from an initially industrial reaction to a variety of ulterior objectives, including both political insurrection and banditry.[15]

Thomis agrees with some of this, but expands on the banditry part. He points out that as the movement developed, it seemed ever more threatening, with all manner of arms raids and other mayhem mixed in with the banditry. He writes, "An almighty crime wave swept over England . . . [breaking] machinery had become a mere excuse for private assassination and robbery. . . . A real crime explosion was detonated by the Luddites, and the masqueraders . . . were debasing the coinage of real Luddism, which was highly motivated and heroically accomplished."[16] The desperation of the un- and underemployed workers also led to some serious food riots, and these too were often pinned on the Luddites, even when they were not involved.

It is clear, however, that the situation was of a magnitude and severity unprecedented in British history, and brought down responses of similarly unusual harshness. In addition to execution, jailing, and exile of those caught, a repressive force of 12,000 military personnel was stationed in the troubled regions.[17] This was a force far larger than that taken by Wellington to Portugal in his battle with Napoleon's troops four years earlier.

Still, the rioting continued till about the end of 1816. There were several reasons for its end. One was the constant and increasing pressure from the government, which included increasing numbers of both troops and spies, leading to more arrests and more executions. But perhaps even more important, the war with France had led to a set of Orders in Council, which had put much of Europe in a state of blockade and had severely restricted trade. These were finally repealed, leading to better times, and were an important factor in an apparently rapid alleviation of the textile workers' poverty.

## *The Importance of the Luddites*

One reason the Luddite story is so important is that it not only involved an industry that played a major role in the rise of England as an economic power but also, in a major way, provided a basis and testing ground for the Industrial Revolution itself. According to the textile historian Edith A. Standen, "The first factories were built to make textiles, the first processes of mechanization were applied to them, and their production and distribution were the first to be organized on a capitalistic basis; the wish to produce them quickly, cheaply, and in enormous quantities was one of the main causes of the Industrial Revolution."[18]

Thompson concurs: "Cotton was certainly the pace-making industry of the Industrial Revolution."[19] The cotton mill was, however, also the model for the image of the "dark, Satanic mill" so grimly depicted by William Blake.[20] By the 1770s and 1780s, there were indeed large mills, employing not only women but very young children for long hours. As early as 1771, in fact, Richard Arkwright and several partners had set up a large, water-powered factory in Cromford, Derbyshire, a town not far from Nottingham. In 1776, his second Cromford Mill was over 120 feet long and seven stories high. By 1782, he employed five thousand workers in his several mills. Eight years later, he installed steam-powered machinery in his Nottingham factory.

But, regardless of the activities of the Luddites and their supporters, and of the supposedly meteoric rise of industrialization, textile manufacturing in England remained largely a cottage industry for decades. Thompson points out: "For half a century after the 'breakthrough' of the cotton-mill (around 1780) the mill workers remained as a minority of the adult labour force in the cotton industry itself. In the early 1830s the cotton hand-loom weavers alone still outnumbered all the men and women in spinning and weaving mills of cotton, wool, and silk combined."[21] The term *manufacture*—making things by hand—still made sense.

One reason the Luddites rose up in England and not, for example, in the United States, had to do with a curious difference between the two countries. As one scholar in the history of technology, D. S. L. Cardwell, explained it, "[I]n America during the nineteenth century land was cheap and labour, especially skilled labor, was expensive. In Britain the opposite conditions prevailed: land was expensive and labour was cheap. Accordingly there was in America a strong incentive to invent and apply labour saving machinery."[22]

Still, some of the early developments in industrialization did arise in England, and certainly played a part in the riots. But the story is not simple. As noted earlier, the population increase and a major upsurge of trade in the eighteenth century had created a strong demand for yarn to be spun for weaving. The inability of the cottage system to produce sufficient yarn provided an impetus for innovations in spinning technology.[23] A good example of such technology is James Hargreaves's spinning jenny (invented 1764, patented 1770), which enabled a number of threads to be spun simultaneously by one person. But it was designed not only to be housed in homes but to be human powered, as well.

On the other hand, an *oversupply* of hand-loom weavers in the wool industry delayed adoption of automated looms and was a basic factor in Luddite rioting in the Yorkshire region.

New inventions did not invariably lead to greater industrialization. One scholar, G. N. von Tunzelman, even argues that use of the spinning jenny in activities such as woolen spinning "actually permitted the domestic system to survive longer. Nevertheless," he adds, "most advances were eventually absorbed into factories."[24]

## *Background of the Luddite Movement*

My own feeling, it is probably clear by now, is that today's technophobes, who use the Luddites as their battle cry and emblem, simply misread the Luddite saga. It was not at heart an attempt to halt the progress of technology. Remember that in spite of the Luddites' largely rural outlook, the operators were well acquainted with technological innovation, and in some cases, profited handsomely from it. For example, the entire industry took off after new, mechanized spinning methods were introduced that sped up the production of cotton yarn in the mid-1700s. The Luddites were both smarter and more realistic than to think that they could hold back continuing mechanization. Except for the croppers—a small, select group—many, perhaps most, of the workers probably were not averse to machines per se.

This view is supported by what was happening in the world of science in their day. Whereas the American and the French Revolutions had stirred up a cauldron of radical ideas in politics, in England any passion for the new was more likely to be satisfied in the world of science.[25] And among its enthusiasts could be found many of the country's artisans, including, no doubt, many of the skilled workers in the textile field.

At the same time and probably connected with this admiration of science, a sea change was taking place in the world of technological innovation. Prior to then, most such innovations were the work of tinkerers; almost any skilled worker could come up with a new way of doing things, or even of creating a major technological innovation.

The story of Richard Arkwright, the powerful owner mentioned earlier, is a good case in point. His training and early experience were as a barber and a wig-maker. Changing fashions in the mid-eighteenth century led to a drop in demand for wigs, however, and Arkwright was looking around for another source of income. He chanced to meet a reed-maker and a clock-maker who were trying to build a cotton spinning machine to answer the need for a major increase in yarn production. Result: Arkwright, who had no formal training in science or engineering, came up with a machine that opened the door to the Industrial Revolution. In fact, he even lacked the ability to construct a model, and John Kay, the clock-maker, built it for him.

The steam engine is another good example. Thomas Newcomen, a blacksmith, came up with the first useful steam, or fire, engine in 1712. Used in mines for many years to pump out unwanted water, it was nevertheless highly inefficient. John Smeaton, a lawyer turned instrument-maker, made improvements in the engine by means of his directed, careful experiments, but it remained for James Watt, a professionally trained instrument-maker who used a far more mathematical, scientific, and theoretical approach, to bring the steam engine to the point where it could be economically employed in textile manufacture. His double-cylinder design dates back to 1765 and so was in existence for decades before the Luddites began their activities.

In general, then, neither scientifically based invention nor technological innovation was new to the Luddites. Clearly, they were mainly interested in getting back at a system that regarded them as little more than chattel, by hitting the owners where it would hurt the most.

## *Effects of the Luddite Rebellion*

Few of us would wish that the Luddites had succeeded in halting the spread of technology. But we can hardly blame them for trying. Still, though the Luddite uprising had little effect on the long-term rise of technology, it has had some very real effects on the society that spawned it. The Luddites so alarmed the authorities that all of what Jacques Ellul calls the techniques of the state—financial, military, police, administrative, and political—were deeply affected.[26] The criminal justice system, for example, long a haphazard hodgepodge, was tightened and strengthened. In addition, some of the acts that so infuriated the Luddites, such as the Combination Laws, were repealed. Finally, the workers learned that by working together they could successfully challenge the hefty forces of authority. In that sense, the political fallout of Luddism was more important than the economic consequences.

Further, the Luddites' powerful image has lived on and even prospered in our own day (see chapter 10). Over the years since then, it has shown up in some surprising ways, especially in the world of literature.

In fact, one of the unquenchable images in our society, one that has lasted virtually undiminished for almost two hundred years, is Frankenstein's monster. (Frankenstein was the name of the doctor who created him, not that of the monster.) Yet few today understand that the monster was, and remains, an allegory of science and technology gone wild. And it came directly out of the Luddite story.

Among the Luddites' defenders was the famous British poet Lord Byron. When Parliament was debating the institution of a death penalty for frame-breaking in 1812, Byron (he really was a Lord) stood up and presented a strong argument against the proposal—using some of the same reasoning the workers had been offering right along. He spoke of "men sacrificed to improvements in mechanism."[27]

Though he failed to carry the day, his feelings about the Luddites and their cause had a powerful, and lasting, effect. This came about in a curious way. He had spent the summer of 1816 cooped up in a Swiss villa during a period of rainy weather. Among his companions were the Shelleys, Mary and Percy Bysshe, and one of the ways they kept themselves occupied was to set up a ghost story competition. Mary, well aware of Byron's feelings about the Luddite movement, personalized them in her story, which she decided to turn into a novel. By December, she was working on chapter 4, in which she details Frankenstein's objective: "I thought, if I could bestow animation upon lifeless matter, I might in process of time . . . renew life where death had apparently devoted the body to corruption."[28]

Sounds like a typical optimistic view of medical science. The title of the resulting novel, *Frankenstein, or the Modern Prometheus,* suggested otherwise, however. And it reflected in a potent and personalized way a fear or warning that there are areas in science and technology where humans should not tread, that new technologies especially hold fearful risks, that in fact the old ways may well be better. Mary Wollstonecraft Shelley, age nineteen, had created one of the enduring images of our modern era.

Throughout the Romantic era, such noted essayists as Ruskin, Carlyle, Emerson, and Thoreau incorporated the idea into their writings, while William Morris incorporated it into his Arts and Crafts movement. It can be seen in the early-twentieth-century Southern Agrarian writings in the United States. A fair amount of science fiction reflects the general idea. Examples include the powerful antiutopia novels of Karel Capek (*R.U.R.* [Rossum's Universal Robots], 1920), Aldous Huxley (*Brave New World,* 1932), George Orwell (*1984,* 1949), and Anthony Burgess (*A Clockwork Orange,* 1962).

A variety of modern writings reflect similar feelings, as for example, those of Lewis Mumford, Jacques Ellul, and Edward Goldsmith. Even Kirkpatrick Sale, who has done a recent history of the Luddites, reflects this attitude. Though each of these writers might, if given the chance, distance himself from a true antitechnology stance, their basic fears shine through clearly in their writings.

It's true that some of these writings, notably such powerful novels as *Brave New World, 1984,* and *A Clockwork Orange,* are really aimed at the totalitarian superstate. But Ellul's point is that modern technology and the power of the state are inextricably entwined.

We often hear that those who forget or ignore the past are condemned to repeat it. But it is also true that those who look back often see what they want to see; they often find what they want to find. So it is with the meaning of the Luddite story.

In any case, neither the Luddite uprising nor its powerful image has presented much of a barrier to the progress of technology. On the other hand, the public and, perhaps to a lesser extent, the scientific establishment itself, have today a less optimistic view of this progress than was common a century ago. Rather, what we are seeing more of is a kind of wary acceptance, a recognition that there are costs as well as benefits and that it is important to try to balance these before permitting a new technology to move forward.

As for the riots themselves, they were, to some extent, the lashing out of a group that was caught in the chaos of a changing industry. But even more, they were a response to the Luddites' economic situation rather than a revolt against technology.

That is, while the economic *results* were less important than the political and administrative fallout, the *causes* of nineteenth-century Luddism were indeed economic. Although some of the feuds I cover in the coming chapters were matters of pride, ambition, competition, and other personal factors, many, as we'll see, also had much to do with money and profit.

# CHAPTER 2

# *Davy versus Stephenson*

## Who Invented the Miner's Safety Lamp?

Even as the Luddites were trying to stave off the inevitable, the world was exploding around them—in more ways than one. On May 25, 1812, the same year the Luddite rebellion was in full swing, ninety men and boys lost their lives in a terrible explosion at the Felling Colliery in northern England. Though it was the worst mine explosion to date in terms of lives lost, it was hardly the first in the coal-mining region in and around Newcastle.

Further, the Industrial Revolution was taking off, and so was its need for more and more power. Man-, horse-, and waterpower had done well for centuries, but now the cry was for coal. By the turn of the nineteenth century, hundreds of Watt's steam engines were already in use, calling for more coal.

But an even greater change was in the works. A new development, the high-pressure steam engine, was not only more powerful but more compact. This made it convenient to use the device, for the first time, to power conveyances. By 1804, Richard Trevithick had put a primitive steam locomotive on rails and used it to pull a line of coal trucks. Railroads shortly became the largest user of coal by far. Coal was, first of all, the prime energy source for the locomotive steam engines. But in addition, rapid expansion of railway lines required not only great quantities of iron, but also lots of coal, needed for smelting the iron.

Early mines had always been shallow. Now, not only were new mines dug but old ones were pushed wider and deeper. As the mines started to go deeper, however, a serious problem arose.

### *A New Hazard*

Though underground mining of all sorts has always been hazardous, coal mining presents special problems. In addition to the constant danger of rock falls and shaft collapses, there is also the ever present danger of fire and explosion. Coal dust, under certain conditions, is explosive, and the same holds for the atmosphere in which the men work. Methane, a gas produced in the conversion of decaying vegetation into coal, often collects in fissures in the mine walls. Widely found in nature, it is the principal component of natural gas. When a miner's pick cuts into a coal seam, some of this gas often will be released. The problem is that the gas is both colorless and odorless, and therefore is undetectable to the miner.

Most feared were the "blowers": a miner's pick would break into a chamber that had been collecting the gas for years or centuries, unleashing a stream of it that might quickly fill the shaft. At concentrations as low as 5 percent, the mixture of gas and air, called firedamp, was already explosive. The combination of firedamp with the miners' open candles or, less often, oil lamps, was a recipe for disaster.

It took a brave soul indeed to be the first in to test the situation. Typically, the gas, being lighter than air, might collect along the roof of a mine tunnel during the night. J. W. Whitaker, a colliery manager and college lecturer, described what happens next in his 1928 book on mine lighting: "Before the miners descended to work in the morning the fireman, wrapped in wet clothes and carrying a long pole on the end of which was set his lighted candle, crept along the roadway and 'tried' the roof for 'gas' until he ignited it, [at which point] he immediately lay flat on the ground to allow the flame to pass over him. Occasionally, instead of a mild inflammation, the ignition of the gas developed into an explosion and caused great havoc."[1]

A graphic description was provided in 1869 by the French writer Simonin: "The moment the mixed gas comes in contact with the flame of a lamp a tremendous explosion takes place. . . . [It] spreads instantly into all the galleries of the mine; a roaring whirlwind of flaming air destroys everything it encounters . . . mounts into the shaft, and lifts from their foundations the staging [building and equipment] which covers its mouth, through which it discharges thick clouds of coal, stone, and timber.

"The men are blinded, thrown down, scorched, and sometimes burnt to a cinder."[2]

Ventilation helped—sometimes. An interesting method consisted of tying a bush to the end of a rope and running it rapidly up and down the shaft. This provided some movement to the contaminated air and, it was hoped, drove it out of the mine workings. Occasionally, a steam engine was connected to some sort of primitive air pump. This helped somewhat, especially at a large mine, but as the results showed, not a great deal.

Even if the initial test showed that no gas was present, as the miners dug and blasted, there was always the fear, and the very real danger, that new gases were being generated. John Buddle pointed out in 1813 that the presence of the gas could be detected by careful monitoring of the open flame. "The first indication of the presence of inflammable air," he wrote, "is a slight tinge of blue, or bluish grey colour, shooting up from the top of the spire of the candle, and terminating in a fine extended point. This spire increases in size, and receives a deeper tinge of blue as it rises through an increased proportion of inflammable gas, till it reaches the firing point; but the experienced collier knows accurately enough all the gradations."[3] Among other specialty workers in the mines, then, was one called the candle-watcher, which doesn't need much explanation.

## *Desperation*

How desperate were the miners to find a solution? One attempt involved use of "phosphoric lights," meaning in this case fish in the beginnings of putrescence. It's not hard to imagine how much light—and smell—that gave off. It wasn't much of a solution.

Between 1806 and 1813, more than three hundred men were killed and thousands more were injured in the coal mines. Yet outside the coal districts, no one paid much attention. Indeed, the coal mine owners did all they could to hush up the tragedies, though they often sustained major losses from such explosions. The negative publicity from such blasts apparently kept the mine owners from reaching out for help, at least at first.

William Reid Clanny (1776–1850), a physician in the area who had seen some of the terrible results of these disasters, began work on the problem in 1811. Although he came up with a lamp that would be safer than an open candle, his design required that the air going through it be driven through a container of water by a hand bellows—meaning that someone had to man this device at all times. The idea was tried but rejected as too unwieldy. In the meantime, however, he wrote a paper on the design, which was read at a meeting of the Royal Society of London in May 1813 and published in the society's *Transactions* in the same year.

With the 1812 explosion at the Felling Colliery, and still others afterward, it was no longer possible to keep this disastrous situation under wraps. The grieving relatives and friends of the dead demanded that something be done. The public cried for action. Politicians wanted results. At that point, even the mine owners sought help.

This was a serious matter. The Royal Society of London had long been interested but had not had any success. In October 1813, a smaller group, which included Clanny, formed the Society for Preventing Accidents in Coal

Mines. It was established in what is now called Sunderland, a small city on the northeast coast, not far from Newcastle. The idea was to raise funds that would be awarded for improved methods of lighting and ventilation.

The publicized requests for help brought some strange answers. One individual suggested filling the mines with chlorine gas. This indeed would have prevented explosions, but it would have poisoned the miners.

Consultations by the Sunderland group with a variety of experts elicited a rather surprising suggestion—for the time. John Buddle, though a ventilation specialist himself, suggested that mechanical means—that is, ventilation—were not likely to prove successful and that the group should look rather to chemical methods for rendering the firedamp harmless. Out of this came the idea that they should turn to scientists for an answer. At a time when an ounce of practice was still thought to be worth a pound of theory, this was actually a revolutionary idea.

## *Sir Humphry Davy*

It was inevitable that the name of Sir Humphry Davy should come up. Davy (1778–1829) not only was a distinguished chemist but had become famous for his wonderful and well-attended lectures on science for the public. He had been appointed to the newly established Royal Institution in London as lecturer in 1801, had become a full professor in 1802—at age twenty-four!—and had been elected a fellow of the prestigious Royal Society of London a year later. In a day when Paris was the center of the scientific world, Davy, almost single-handedly, brought England back into the limelight. He was knighted for his scientific work in 1812.

By then he had studied not only chemistry but also geology, biology, and physics on his own, and had done useful work in these areas. He had spent time working at the Pneumatic Institution, studying the physiological effects of various gases in the treatment of illness; he had discovered the first chemical anesthetic (nitrous oxide, still sometimes used in dentistry); and, by July 1800, at the age of twenty-one, he had published a six-hundred-page tome, *Researches, Chemical and Philosophical; Chiefly Concerning Nitrous Oxide and Its Respiration.*

Perhaps most important to his scientific fame, he had done pioneering work on the effect of electricity on chemical compounds, which led to his discovery of a number of important elements, including sodium, potassium, calcium, and magnesium.

Early on, while working at the Pneumatic Institution, he had stayed at the home of its director, the well-known and highly respected Dr. Thomas

Beddoes. There he met and hobnobbed with all kinds of eminent, talented, and impressive people, from poets on the order of Coleridge and Wordsworth to engineer/inventors like James Watt.[4] As the lecturer at the Royal Institution he interacted with scientists on the order of Henry Cavendish, Count Rumford, and Sir Joseph Banks. It was during all of these interchanges that he honed his brilliant wit and speaking ability, which were both to prove so important in his later career.

So illustrious had Davy become that Mary Shelley, who had met him in London, used him as the model for her fictional Professor Waldman, Frankenstein's teacher in her famous novel.

But Davy's fame as a lecturer rested on more than his scientific accomplishments and his speaking abilities. As a kind of standard-bearer for the world of science in that day, he had a vision, and it was one that struck a chord in his illustrious attendees. He presented, in glowing terms, his belief that new scientific developments would help in the betterment of the world. In his words: "We do not look to distant ages, or amuse ourselves with brilliant, though delusive dreams, concerning the infinite improveability of man, the annihilation of labour, disease, and even death. But we reason by analogy from simple facts. We consider only a state of human progression arising out of its present condition. We look for a time that we may reasonably expect, for a bright day of which we already behold the dawn."[5]

He felt that chemistry, thanks to the discoveries of new gases and previously unknown elements and also to its connections with the processes of life, promised to eliminate pain and improve life for everyone.[6]

Remember too that in Davy's day, the lines between the arts, humanities, and science were not nearly so precisely drawn as they are today. Nature and what we would call science today were being looked at with great interest by the Idealist/Romantic poets like Coleridge and Shelley, as well as by scientists like Davy. Goethe, a contemporary, thought of himself as a scientist as well as a poet and a dramatist; Davy thought of himself as a poet as well as a scientist. Though Davy's poetry today is considered fairly pedestrian, both Coleridge and Robert Southey praised it. Davy even corrected the proofs of an edition of the *Lyrical Ballads,* a collaborative work of Wordsworth and Coleridge. Davy was, in other words, the Renaissance man of the day.

Ironically, however, Davy's meteoric rise to fame led him to resign from the Royal Institution in 1813—at the ripe old age of thirty-five. To placate his wife, who was feeling rather left out of his life, he set out with her on a lengthy set of travels in Europe.

With e-mail and cell phones far in the future, Davy was not contacted until the summer of 1815. A letter from the Reverend Dr. Gray, a member of the Society for Preventing Accidents in Coal Mines, finally reached him in Scotland, where he was on a shooting excursion.

To the written invitation to participate in the Sunderland group's guest, he replied on August 3: "It will give me great satisfaction if my chemical knowledge can be of any use in an inquiry so interesting to humanity."[7] On his way back he stopped off at Newcastle, where he visited the mines, conferred with John Buddle, and collected some of the treacherous gas to experiment with.

Davy immediately began looking into the range of concentrations of the firedamp in air at which the mixture becomes explosive and on the degree of heat needed to ignite it. He also looked into whether, and how, a flame would move through apertures of various sizes to an explosive mixture. He found that firedamp will not explode in a small tube if its diameter is less than one-eighth of an inch, or even in a larger tube if there is a mechanical force driving the gas through the tube, or if the tube is of sufficient length. Nor would it pass through a fine wire mesh. The finer the mesh, the safer the lamp—and, unfortunately, the less light it allows through. Nevertheless, it was this work that eventually led to a solution.

Once Davy began work on the problem, it took him less than three months to solve it. By October 30, 1815, he had sent his first report on his studies to the society at Sunderland, based on his researches and on some laboratory experiments. On November 9, he read the paper at a meeting of the prestigious Royal Society of London. In it he outlined his progress, including some findings on the conditions under which firedamp explodes, and how it behaves when mixed with air. It also included his first public mention of introducing air into the lamp through a fine wire gauze.

He gave a second paper on January 11, 1816. In it he presented the idea of a lamp fully enclosed by a wire gauze cylinder. And on January 25, he read a third paper to the society. Here he reported that two of his lamps had been tested in two of the most dangerous mines near Newcastle and had performed successfully. The papers were published later on, as was one additional paper on the method.

So there it was, out in full view of the public. His first lamps were put into service at the beginning of 1816; by the end of that year, they were fairly widely used in the coal fields in northern England. And at least initially, they seemed the answer to a prayer. Davy was feted, given prizes, and widely honored. One letter of gratitude from the miners at Whitehaven Collieries was especially gratifying. Thanking him for his "invaluable discovery of the safe lamps, which are to us life preservers," it was signed by eighty-two miners.[8] Forty-seven, unable to manage their own signatures, signed by putting crosses next to their names.

The long hiatus between the Sunderland group's initial suggestion and their contact with Davy, however, was unfortunate. Had Davy started working earlier on the problem, it might have prevented what turned into a major, and very ugly, priority dispute. But in the time that had passed between the initial suggestion that Davy be consulted and Gray's contact with him, someone else—someone not nearly as famous as Davy—also had looked into the problem, and also had come up with a solution to it.

## *George Stephenson*

George Stephenson (1781–1848) was a nobody in comparison with Davy, at least at that time. Life was difficult for the workers in the coal-mining region around Newcastle, and Stephenson came from a poor family in which none of the six children received any education. At the age of eighteen, he still could not read or write. He learned to do both in later years, but never with facility.

He did some farming in his young years and held a variety of odd jobs, but was happiest when tinkering with anything mechanical. He tended some steam engines in his teens and developed abilities that were to be very useful later on. In 1798, at the age of seventeen, he was put in charge of a pumping engine at a mine called Water Row.

By 1801, he was a brakesman at the Dolly Pit, in charge of a winding engine, which raised the loaded coal cars to the surface and also ferried the men into and out of the pit.

Stephenson married in 1802 and, ambitious, began to study the basics of mechanics on his own. He later, in 1822, paid for his son Robert to take a short course at Edinburgh University. The course covered natural philosophy, natural history, and chemistry. Academic engineering courses, interestingly, still lay several decades in the future. It was a good investment, nevertheless, for Robert was able to share much of what he learned with his father; but it came too late to be of any use to George in his coming battle with Davy.

At home, George did watch and clock repair as a hobby, and his home began to fill with small inventions and engines; he became known as Geordie Stephenson, the engine doctor.

Then, due to some reverses and deaths in the family, George actually considered emigrating to the United States. One such reverse was a disastrous run-in at a parliamentary commission where he gave evidence for the Liverpool and Manchester Railway. It was an unfortunate case of a smart but uneducated man being made to look very foolish by far more experienced and sophisticated government lawyers. It's interesting to speculate on what the course of the Industrial Revolution might have been had Stephenson emigrated for, as we'll see, he played a major part in its development.

In his early years, he was "a wiry, muscular youth who excelled in weight lifting, wrestling, leaping and hammer throwing and who proved, in his celebrated fight with Ned Nelson, the battling pitman of Black Callerton, that he was a match for any man."[9]

By 1806, he had become the brakesman at the Killingworth Colliery, a more responsible position. Located in the spread-out coal mining region that included the counties of Northumberland and Durham, and with almost 160 *miles* of gallery excavation, the Killingworth was undoubtedly the largest mine in that area. It was owned by a rich and influential group called the Grand Allies.

In that same year, an explosion took place in the mine, killing ten men and creating in Stephenson a powerful desire to do something about the problem. A few years later, in 1809, twelve more miners were killed in another explosion; and as we saw at the beginning of this chapter, the explosion at Felling took the lives of ninety miners in 1812. By this time, however, Stephenson was moving along in his career and was already deeply involved in the work that was, later on, to make him as famous as Davy, though in a very different sphere.

His fortunes began to change in 1811. A pumping arrangement at Killingworth, unable to pump enough water, had permitted a deep pit to flood. Recognizing its weaknesses, Stephenson offered to fix the pump. The owners figured they had nothing to lose, and gave him a chance. Fix it he did, with some clever adaptations to the apparatus and changes in the engine itself. For this truly useful and important work, he received the grand sum of ten pounds. But more important, his abilities were starting to be recognized.

By 1814, he had risen to the office of colliery engine-wright. As such, he was personally involved in all the day-to-day operations, which included supervising the carving out of the inclined tunnels through which the coal was brought to the pit entrance. Stephenson already was thinking about the rails over which the mine cars were being drawn. He came up with an improved way of joining the track portions. And thinking even further ahead, he was working on a design for what eventually became the modern railway locomotive, an invention that truly changed the world.

Though others had done some preparatory work along these lines, each of the attempts had severe weaknesses that ensured its failure, and Stephenson later was to be called the father of the locomotive. Stephenson, in fact, not only came up with a workable design, but seemed most able to foresee its incredible future. This, in spite of the fact that the earliest devices were no less temperamental than the horses they would eventually replace. And Stephenson himself reflected his great vision: He told a friend at Killingworth, "I will do something in coming time which will astonish all England."[10] By 1815, he had already gotten several patents on various aspects of the newly developing technology.

His first tangible accomplishment in this field came about when Sir Thomas Liddell asked Stephenson to supervise the construction of a locomotive for use at the Killingworth mine. The result was the first workable steam-driven, locomotive-drawn train over iron tracks. He named it the Blucher, no doubt after the Blucher pit located near the Killingworth mine. The Blucher crawled slowly past his cottage door on July 25, 1814.

The tie between mining and the locomotive was a tight one. As the British historian Eric Hobsbawm put it, "[T]he railway is the child of the mine."[11] And for years, just about all the locomotive drivers came from Stephenson's native coalfield.

Stephenson's work in this area brought his name to the attention of several influential men in the region in which he was working. One of these, William Losh, a senior partner in a major ironworks firm, was to become a major supporter of Stephenson in the coming safety lamp battle.

Though Stephenson had never actually been a miner, he had spent countless hours down below. He was familiar with all aspects of the operations, including the dangers, having himself played a part in helping snuff out a serious threat. Could nothing be done to prevent these all-too-common fires and explosions? It was a question on everyone's mind.

What was needed, it seemed clear, was either some chemical means to negate the inflammable/explosive nature of firedamp, or more likely, a lamp that would give off enough light for the miners to work without igniting any firedamp that might be seeping or even blowing. For though the gas could be explosive, it was not poisonous. Clanny's attempt had already been discarded. What now?

## *"Burnt Air"*

Although Stephenson had been deeply involved in the newly developing world of the railroad, the mine explosions continued to haunt him, and for several years he had been experimenting with firedamp. In his untutored but uncannily intuitive way, he had noted that lit candles would be extinguished if placed downwind of burning blowers. Though he probably didn't understand the chemistry, the burnt air had had much of its oxygen used up and would therefore not support further combustion.

He carried out further experiments, among them bringing lighted candles to the fissure from which gas was escaping into the mine. During these experiments, the miners would do their best to get as far away from him as possible before he lit the gas. One of his theories was that making the velocity of the firedamp/air mixture entering below the lamp flame fast enough would prevent any explosion from moving downward in the lamp, and the burnt air would prevent it from moving upward.

By the summer of 1815, he felt it was time to put these experiments to work, and he began serious work on the problem. He had drawings prepared by August, and by October 21, he had the first lamp ready for testing. As a hands-on type, he chose to test it himself, in a seam that featured a strong blower—one so strong, in fact, that he could hear the hiss. Two colleagues went down with him to witness the experiment, though they stayed well back. Stephenson advanced slowly, holding his lamp in front of him.

Utterly foolhardy? Not really. He figured that no one would believe him if he did not carry out the experiment under real conditions. He also had enough confidence in himself and his lamp to do so. As he proceeded, he watched the flame carefully. At first it grew in size and changed color to a shade of blue.

After a tense wait, his witnesses, waiting back at the entrance, heard Stephenson shout that all was well, but that the lamp had gone out. He had to find his way back from the darkness by groping the walls. A second attempt with the same lamp produced the same result.

He carried out experiment after experiment. His basic idea was that the gas would not ignite if passed through tubes of small diameter before it contacted the flame. In his first design, he had a single tube through which the gas passed. After much experimentation—done in his spare time—he changed this to three small-diameter tubes and then to another in which many tubes were used. But it took a while before he realized that it was not the length of the tubes that mattered, but their small diameter.

By November 30, 1815, he had built and tested his third and final design. It was based on this idea of small apertures. The lamp and its glass chimney were surrounded by a metal plate perforated with small holes.

Davy, in the meantime, also had come up with his design—which used wire gauze instead of perforated metal plate and which did not include a glass chimney, as required in Stephenson's design. Davy had read his first paper to the Royal Society on November 9, a couple of weeks earlier.

The stage was set for some serious combat.

## *An Uneven Contest*

Davy's supporters, wanting to show their appreciation, planned a public demonstration in his honor, but it was put on hold briefly when a question was raised about the possibility of Stephenson's priority. Although the objection was voted down, the wording of the honor was later changed from "the invention of his safety lamp" to "*his* invention of *the* safety lamp."[12]

Stephenson was hampered in the presentation of his case in that he was nowhere near the speaker that Davy was. He never could get rid of an accent that was so heavy as to be almost unintelligible to those living in the southern part of the country. One of Stephenson's biographers, Samuel Smiles, describes him as "so diffident in manner and unpracticed in speech" that when invited by the Philosophical and Literary Society of Newcastle to present his side of the story in December 1815, he took with him his friend Nicholas Wood to act as his expositor. But as the evening proceeded, Stephenson, not fully happy with the way things were going, proceeded to take over, and in his strong Northumbrian accent managed to get his ideas across.

Stephenson also managed to get his side of the debate published. This appeared in the form of a letter to the *Philosophical Magazine*, but not until March 1817. The editor had earlier offered his opinion that Stephenson's ideas on small apertures and safety tubes had been borrowed from Davy's design, reflecting a feeling that was held by others as well.

Stephenson was fighting mad. He argued (a little awkwardly) in his published answer:

> If firedamp were admitted to the flame of a lamp through a small tube,—that it would be consumed by combustion, and that explosion would not pass and communicate with the external gas, was the idea I had embraced as the principle on which a safety lamp might be constructed, and this I stated to several persons long before Sir H. Davy came into this part of the country. The plan of such a lamp was seen by several, and the lamp itself was in the hands of the manufacturer during the time he was here. . . .That I pursued the principle thus discovered and applied, and constructed a lamp with three tubes, and one with small perforations, without knowing that Sir Humphry Davy had adopted the same idea, and without receiving any hint of his experiments, is what I solemnly assert.[13]

All to no avail. In the fall of 1817, Davy was honored and presented with an opulent gold-plated dinner service. Estimates of its value range between £1,500 and £2,000. Today, that would be the equivalent of tens of thousands of dollars.

The awards committee was aware of Stephenson's work, but, says L. T. C. Rolt in his biography of Stephenson, "they contented themselves by voting him a purse of a hundred guineas [just over £100] as a kind of consolation prize for what they regarded as the clumsy efforts of an uneducated man."[14] Many of Davy's supporters, in fact, expressed great indignation at the "presumption" of the fellow. This, in spite of the fact that he had already constructed his first locomotive. His fame, however, took a long time to burst the bounds of his district.

Stephenson's unhappy supporters later raised a separate subscription of £1,000, part of which went toward the purchase of a handsome silver tankard. The award was presented at a public dinner on January 12, 1818.

He also had to make his first major speech in acknowledgment of his award. Brave as he was in other ways, he found this very difficult, and he prepared carefully for it. Surely his listeners, coming mainly from the same area, were able to understand him. But, says Rolt, he had set down on paper what he was going to say, and Rolt quotes an unidentified person who saw the notes as reporting. "[I]t is so ill spelt and lacking in punctuation that its effective delivery is difficult to conceive."[15]

## *The Scientist versus the Engineer*

Both men could have let the matter rest at that point, but it wasn't to be. Davy, no less smug than his supporters, simply could not believe that an unknown, uneducated man possibly could have beaten him out and assumed there had been some sort of chicanery on Stephenson's part. Davy even had the temerity to fire off letters to Stephenson's most prominent supporters. The one to William Losh is worth quoting:

> Sir,
>
> Having seen your name in the papers connected with an opinion which every Man of Science in the Kingdom knows to be false in substance as it is absurd in expression, I wish to know if it is used with your consent.
>
> The Public Scientific Bodies to which I belong must take Cognizance of this indirect attack on my Scientific fame, my honour and varasity (sic). I wish to know my enemies on this occasion.[16]

Losh, to his credit, had no qualms about putting Davy in his place. After some brief introductory material, he got to the point: "Satisfied as I am with my conduct on this subject I must say that I am wholly indifferent as to the cognizance which may be taken of it by the 'Public Scientific Bodies' to which you belong."

Another recipient, the Earl of Strathmore, was even more direct in his reply to Davy. First, he pointed out that Stephenson's was the only safety lamp that had been used at Killingworth and other collieries "in which I am concerned . . . The men who work in them are perfectly satisfied with those lamps and no explosion has taken place in any of our collieries since their introduction."

Further on, he added, "No man can more highly appreciate your merits than I do . . . but at the same time I can never allow any meritorious Individual to be cried down because he happens to be placed in an obscure situation—on the contrary, that very circumstance will operate in me as an additional stimulus to endeavour to protect him against all overbearing efforts."

Local patriotism played a part here. Davy was a Cornishman turned Londoner; Stephenson's supporters saw him as a "distant expert" and therefore automatically suspect.[17] They found it hard to believe that someone who was without Stephenson's extensive coal-mining background—and who undoubtedly was a head-in-the-clouds-scientist to boot—could possibly have come up with the solution on his own.

Ironically, Davy was far from being a head-in-the-clouds scientist. He was an excellent experimentalist, and in his way, he probably took as many chances as Stephenson. He nearly killed himself by seeing what would happen if he inhaled carbon monoxide, which is now known to be a deadly poison, and he sustained a number of injuries during his many experiments with inflammable and explosive materials. And although he carried out his early experiments in a laboratory, he eventually did test his lamps under actual field conditions.

Accusations and rumors that each had stolen the idea from the other—that is, had gotten it via some underhanded method from perhaps an unsuspecting friend or colleague of either man—passed back and forth, to little effect.

Someone started a rumor, for example, that Davy had earlier visited a mine where Stephenson's lamp was already in use. Davy was incensed and insisted that he had never even heard of Stephenson until weeks after he had already published his first paper.

Oddly, Davy expressed his indignation in private, never referring to Stephenson in public—perhaps because he would have had to admit that Stephenson did present a lamp before he did. But his indignation was real and strong. In one letter, he used the term *infamous* to describe the suggestions of Stephenson's supporters that he had pirated the invention. "It will turn out," he wrote, "a very disgraceful business for the persons who have agitated it." In another letter, he wrote that "there never was a more gross imposture than that of Stephenson."[18]

The conflict continued to reverberate, even after the two men backed away from it. When Smiles published his biography of Stephenson in 1859, for example, he made some strong statements. He stated flatly that Stephenson had the basic work before Davy, and that while it was by no means the complete answer, it was certainly enough and that Stephenson should be given the credit.

"It is true," he admitted, that Stephenson's "theory of the 'burnt air,' and of 'the draught,' were wrong; but," he argued, "his lamp was right." Smiles compared the situation with that of two seventeenth-century physicists: "Torricelli did not know the rationale of his Tube [mercury barometer], nor Otto Guericke that of his Air-pump. Yet no one thinks of denying them the merit of their inventions on that account."[19]

Obviously, we must take into account Smiles's quite apparent bias. The story is nowhere near as one-sided as he pictured it. The most likely explanation is that both men had come up with the idea truly independently. While simultaneous discovery seems amazing at first, it happens more often than one would expect. Other examples include Faraday and Henry (electromagnetic induction), Adams and Leverrier (discovery of Neptune), Darwin and Wallace (theory of evolution), Heisenberg and Schroedinger (quantum mechanics), and Schally and Guillemin (thyroid-stimulating hormone).

It's worth noting that the basic idea of the tubes had been advanced some years earlier by Smithson Tennant and William Hyde Wollaston. But they made no mention of its possible use in safety lamps, and it's also quite possible that neither Stephenson nor Davy had learned of this work prior to their own.

## *The Lamps*

In any case, Davy's fame made it inevitable that his work would be more widely recognized and applied. The Davy lamp became the standard in many mines; hundreds were used, whereas Stephenson's were used mostly within his home territory. Davy's design did have certain advantages. He showed that the tubes through which the gases were drawn could be shorter than the ones Stephenson put in. Also, he constructed the lamp with only a wire gauze around the lamp, without the additional glass enclosure that Stephenson's called for. This permitted a simplification of the design as well as the production of more light. He also developed some of the basic theory that provided a foundation for further development later on.

It is to both men's credit that neither tried to patent his invention. Both wanted their lamps to be used for the miners' good and asked no monetary reward for them. On the other hand, both men were strong-minded, independent, and proud, and both apparently wanted the credit so badly that they ended up fighting bitterly about who was to get it.

There is a powerful irony here. In spite of all the honor, the credit, the prizes—to both, but mainly to Davy—the long-term result was confusing. Over the long haul, explosions and deaths did not decline. Consider the abundant and extensive coal mines of Durham and Northumberland. In the eighteen years prior to introduction of the Davy lamp, 447 miners lost their lives. In the eighteen years following, the number rose to 538.[20]

One explanation for these forbidding figures is that the shafts and tunnels were dug even deeper than before, and mines that had been closed because of the dangers were reopened, exposing the miners to even more dangerous conditions. Second, the miners, thinking they were now safe, may have become somewhat more careless. Finally, the wide use of the lamps (mainly Davy's design) also contributed, for the lamps were pushed to their extreme: the wire gauze that Davy suggested and which was wrapped around the lamps could become clogged with dust and also become red-hot under certain conditions, which may well have led to unexpected explosions.

Note, however, that far more coal was being produced, so the number of disasters per unit of coal produced may well have decreased.

In any case, neither design was the final answer. While Stephenson's lamp provided less light, and was also heavier and more expensive, it was somewhat safer. David Knight, a Davy scholar, points out that Davy's laboratory studies were done with samples of gas in which any coal dust would have settled out. Stephenson's in situ studies showed more clearly the additional danger of coal dust in mine air.[21]

At the Oaks Colliery in Barnsley, for example, both types of lamp were being used. A sudden influx of firedamp caused the Davy lamps to become red-hot, whereas all the Geordie lamps went out. Happily, there was no explosion. But in a different case, at the Gosforth Colliery at Barnsley in January 1825, when the top of a Davy lamp became red-hot, it did indeed cause an explosion that led to the death of two dozen men and boys. On the other hand, experience showed that when exposed to strong air currents, Stephenson's lamp could cause explosions as well.

## *Later Safety Lamps*

In 1835, when it was clear that the Davy lamp, though most widely used, had certainly not been the final answer, J. J. Wilkinson, a lawyer and founding member of the Sunderland group, pointed out that it "suffered from the indiscreet zeal of his friends in not allowing it to be an imperfect instrument."[22]

Still, Davy's work, which looked into the theoretical underpinnings of his design as well as the principles of safety, helped lead to safer lamps. One incarnation, introduced in 1833, was actually a combination of the Davy and Stephenson designs. But again, it was not widely adapted because of its low light output. Clanny, too, came up with another design. It was essentially a Davy lamp with the gauze around the flame replaced by reinforced glass and using a different air feed setup. It produced more light, and found some use. But if the velocity of air moving through the lamp exceeded seven feet per second, it lost its safety features.

That none of the lamps was the final answer was clear, and the Society for Preventing Accidents in Coal Mines was still in existence three decades later, with Clanny—who had continued to work on the problem—as secretary and treasurer.

Clanny's contributions had remained buried for years under the conflagration raging around the Davy-Stephenson conflict. And as if the priority battle were not confusing enough, Clanny began to argue loudly that *he* was the one who had been robbed, that he should have been recognized as the inventor of the flame safety lamp. There were those who agreed with him, and on February 3, 1848, he was presented with a splendid silver tray and a purse of gold at one of the Sunderland society's meetings. It came none too soon, for he died two years later.

As far as I can determine, Davy never publicly recognized Clanny's contributions. Stephenson, on the other hand, actually contributed a few pounds to the public subscription for Clanny. But he made sure to point out that Clanny's lamp had not proved practicable to manage in coal mines.[23]

Various configurations of flame safety lamps continued to be developed and widely used right into the twentieth century. A British commission of 1886 investigated 250 different configurations. Each addition, however, although beneficial, added its own problems.

Still, the flame-based lamp continued to develop. A 1924 report by the U.S. Department of Mines stated that the lamps by that time were providing five to ten times as much light as the original designs, and were far safer when used by the average miner.[24]

A move to electricity was the next logical step, and once the switchover started, it moved fast. For example, in 1911, all lamps in the bituminous mines of Pennsylvania were of the flame type. Seven years later, there were 48,000 electric lamps and only 17,000 flame lamps. The problem with electrical lamps, however, is that they give no indication of any lurking dangers from firedamp, which the flame of the flame safety lamp does offer.

There seems in fact not to be a final answer. Coal mining remains a dangerous occupation, as attested to by recent explosions all around the world. Examples include one in a mine in Ukraine in August 2001, in which dozens of miners died and many more were reported missing,[25] and two gas explosions in Brookwood, Alabama, in September 2001, in which four miners were known dead and nine of their attempted rescuers were missing at the time of the report.[26] A Web site devoted to the problem, moles.org, reported that just in the month of March 2001, 213 coal miners died in China as a direct result of gas explosions.

## *Matters of Class*

While all cases of simultaneous discovery offer the opportunity for priority conflict, there were in this case some special factors. Stephenson, like Davy, was not what we would call a modest man. Though not yet recognized by the rest of the world, he nevertheless had a high opinion of himself and his abilities. Rolt refers to "the flaws in George Stephenson's character which became increasingly evident as he rose to fame. He was a proud and jealous man who would acknowledge no peer and brook no contradiction in the field he made his own."[27] Rolt was referring to the railroad business, but with the small exception of his contribution to the Clanny subscription, the description undoubtedly applies in the safety lamp story as well.

Further, although Stephenson had an uncanny ability to deal intelligently and with deep insight with anything mechanical, Rolt says he "affected to despise scholars and theorists and he would maintain his views against theirs, right or wrong, with a stubbornness which could be pig-headed." Rolt adds that Stephenson felt his "complete lack of scholarship and formal engineering training very keenly, covering with a cloak of arrogance this sense of inferiority. These characteristics explain the difficulties, the conflicts and the jealousies which marked and marred his astonishing career."[28] His row with Davy was a major example, though there were others.

Although Stephenson went on to greater glory after the first safety lamp days, the class card, so important in nineteenth-century England, had a strong effect on him. Rolt writes, "The London engineers for long regarded Stephenson as an impostor and it must be acknowledged that by their professional standards he never possessed any technical qualifications as a civil engineer." And so he never fitted comfortably into that esteemed group, even though their interests included canal and turnpike building, as well as other aspects of transportation. Nor did he become a Fellow of the Royal Society. He later maintained, however, that he had been asked and had declined in both cases, not being interested in "these empty additions to my name."[29]

It's not hard to understand why Stephenson felt he needed the credit for inventing the safety lamp. He was trying to build a reputation, one that would match his own elevated feelings about himself. Further, as the social underdog, he was fighting a losing battle.

But things are never as simple as they may seem at first. So, when we wonder why Davy fought so hard, we come up with an interesting fact. Davy might well have had similar feelings of social inferiority. Although he was brilliant and famous, and moved in a respected social circle, he still had to face the fact that he had been born into a working class family and, though widely read, was basically self-taught. Thus, he was simply not a member of the old-boy class. Though in our day, such a "serious" shortcoming would hardly be held against him, it was a factor in the upper circles of his day, when class mattered very much.

He may therefore also have suffered from a different kind of snobbishness in that, no doubt because of his lack of academic scientific training and his own proclivities, he was in no way the still-widely-admired pure theoretician. A good experimenter, he actually felt that the only value of hypotheses was that they led to new experiments. In a day when experimental science was still something of a newcomer, this too could have given him a bit of an inferiority complex—though undoubtedly, just a bit—and caused a continuing need to prove himself.

As an indication of the type of man he was, Davy in 1824 tried to block Michael Faraday's election to the Royal Society even though he had been Davy's assistant there for several years. One reason may be procedural: Faraday's application was posted without Davy's knowledge. He undoubtedly wanted to put up Faraday himself, when *he* thought Faraday was ready.[30] Fortunately, his attempted block did not succeed.

After Davy's death in 1829, Faraday was asked about his first meeting with Davy. His answer shows that Davy recognized even then the clay feet of lofty creatures, perhaps including his own: "He smiled at my notion of the superior moral feelings of philosophic men," Faraday wrote in a letter, "and said he would leave me to the experience of a few years to set me right on that matter."[31]

## *Last Days*

Regardless of Stephenson's class problems, he was celebrated by both the public and the press. He also had the satisfaction of seeing his son Robert rake in honors of his own, including accession to vice president of the Institution of Civil Engineers in 1847 (moving up to president in 1857). This, says Rolt, was in spite of George's "obstinacy, his egotism and his intransigent temper [which] had shadowed the Stephenson name with dissension and enmity." Robert managed this "by a scrupulous avoidance of his father's mistakes."[32]

Robert also, it should be noted, lacked his father's enormous self-confidence, which had helped George overcome obstacles that might have stymied a lesser man. By George's death on August 12, 1848, his contributions had been well recognized; his funeral was well attended—though it was more likely to be due to respect for his accomplishments than love for the man.

Sadly, it appears that in spite of his later successes, the feud with Davy had hung heavily on his mind, and he died a bitter and unhappy man.[33]

Davy's last days, too, were difficult ones. Although he had been president of the esteemed Royal Society since 1820, there were complaints about his stewardship—including his opposition to Faraday's election as a Fellow of the society.

When the Royal Society was asked to look into the problem of corrosion in the copper sheathing commonly used on ship bottoms, he took on the research himself and came up with the idea of rendering the sheeting electrically negative. Unfortunately, although the corrosion was largely corrected, marine organisms now adhered so tenaciously to the protected metal that the ships were worse off than before. This provided more ammunition for his enemies, who saw him as "arrogant and high-handed."[34] Even the press, ever fickle, went after him, ridiculing both him and his idea.

Some of his earlier admirers also became disillusioned over time. Coleridge, for example, earlier had felt that Davy's ideas were more "ennobling" than Newton's, but later was annoyed by what he saw as Davy's pretentiousness. He also, suggests Kipperman and others, looked at Davy's safety lamp as yet another example of technology gone wrong. Rather than saving lives, thought Coleridge, it simply permitted mines to be dug deeper and more lives to be lost. In any case, he was distinctly annoyed when Davy was knighted on April 8, 1812. It was one day later that Davy gave his farewell lecture at the Royal Institution. Among those attending and taking notes was a young, budding scientist by the name of Michael Faraday.

In 1826, Davy suffered a stroke. It was not a major one, and he was able to move about Europe on a variety of lonely journeys, seeking an improvement in his health. He also did some hunting and fishing, both of which he loved. But his health was failing, and he died on May 28, 1829.

In spite of all the honors heaped upon Davy, Knight tells us, he "grumbled at the end of his life that he had not had sufficient honor and recognition."[35]

So we have two very brilliant, accomplished, and honored human beings who could have had the world at their fingertips. Yet they left an aftertaste of unhappiness, jealousy, and grousing.

On the other hand, we should keep in mind that while the confluence of their work led first to controversy, it resulted eventually in an improved, and important, product. In his 1966 biography of Davy, Sir Harold Hartley declared flatly, "No other industrial invention can compare with [the flame safety lamp] in the saving of lives and human suffering, and without it the great expansion of coal mining would have been impossible."[36]

Further, all pioneering work in technology requires both theory and practice. So whoever gets the credit shouldn't matter, as far as posterity is concerned. We can only be thankful that there are people who are not only capable but willing to devote themselves to a problem plaguing their fellows, enough to come up with an answer.

# CHAPTER 3

# *Morse versus Jackson and Henry*

## The Electromagnetic Telegraph

Samuel F. B. Morse was not the first to come up with a telegraph alphabet code; he did not invent the electromagnet; he was not the first to build an electric telegraph; he was not even the first to think of one. Did he have the right to call himself the inventor of the electromagnetic telegraph, and to get rich from its profits?

He and some followers thought so. A lot of people thought not. So although Morse's telegraph rose to the top in the world marketplace, and was even deemed patentable by the United States Patent Office, a surprising mix of inventors, scientists, and entrepreneurs kept rising up, claiming to have built a working telegraph before he did. One opponent was later able to create a list of sixty-two claimants![1]

Others argued that they had contributed so significantly to Morse's design that they should share in the profits that began, eventually, to pour in. Though Morse continued to prevail in one controversy after another, his life for years was one of constant struggle and turmoil.

### *An Active Mind*

Until 1832, when Morse turned forty-one, he had been an artist. Fairly well known, he was respected by his colleagues and had received some important commissions. He helped found the National Academy of Design in New York, and even served as its president from 1827 to 1845. But somehow his grand plans for an art career never seemed to work out. Commissions for historical paintings, which he loved doing, were few and far between; so he turned to portrait painting but found himself struggling there as well. Life for the artist then was no easier than it is now.

Morse, however, had an extraordinarily active and fertile mind. Though he majored in art during his student days at Yale from 1807 to 1810, his wide range of interests drew him to lectures and even lab work in the newly developing world of electricity. Among the lecturers were Dr. Jeremiah T. Day, who went on to become president of the university, and Dr. Benjamin Silliman, who became one of the country's best-known scientists.

Later, during slow periods in Morse's art career—and there were many—his studio became a laboratory where he experimented with theories of color and with variations in pigments, oils, and varnishes. He also came up with an important improvement in the common water pump, which later translated into an improved pumping engine for fire trucks. As with his artwork, however, these inventions brought accolades but little income.

Still, he managed to spend three years in Europe, from 1829 to 1832, studying and refining his technique. While there, he witnessed and was deeply impressed by what was then the most up-to-date and fastest mode of communication, the optical semaphore. With this system, a message could be sent from, say, Lille to Paris (144 miles) in a couple of minutes. Developed by Claude Chappe, a French engineer, at the end of the eighteenth century, it was similar to the flag-based semaphore system still sometimes used by railroads and ships today. In tall platforms spaced five to ten miles apart, operators manipulated wooden arms, the varied positions of which represented specific letters and numbers. The system had serious limitations, including blackout periods during rain, snow, or fog and after nightfall. But in an age when the fastest alternative was still a horse's back, it was impressive.

Morse loved his stay in Europe, but eventually it came time to head back to the United States.

## *A Momentous Trip*

The packet-ship *Sully,* captained by William Pell, was scheduled to sail from Havre to New York on October 1, 1832, but stormy weather kept the ship harborbound. Five tedious days later, it finally left port.

During these long trips across the Atlantic, which commonly spanned weeks and sometimes months, conversation and reading were the main forms of entertainment. New discoveries in the world of electrical science provided some fascinating subject matter for such conversations.

Think of it: Only two decades earlier, Hans Christian Oersted, a Danish physicist, had shown for the first time that electricity and magnetism were closely related. A new science, electromagnetism, had arisen. Then the French physicist André Marie Ampère found that an electrified coil of wire behaves like a bar magnet. Based on that development, the British physicist William Sturgeon built a horseshoe-shaped bar of iron surrounded by such a coil and showed that it could lift twenty times its own weight. Thus was born the electromagnet, perhaps the most fundamental electrical instrument ever devised, and one that Morse later put to good use. New developments continued to emerge.

As a result, the subject of electricity was being discussed everywhere. The good ship *Sully* was no exception, though not one among the passengers or crew was a scientist or engineer as we would know those titles today.

Two of the passengers, however, came close. Morse, as we've seen, had a long-standing interest in the subject. And Dr. Charles T. Jackson, though a physician from Boston, had become fascinated by these new developments and was just returning from Paris, where he had learned more about them from some of the top people in the field.

During one shipboard discussion, a member of the group wondered aloud whether distance slows down the movement of electricity through a wire. Jackson referred to experiments by Benjamin Franklin, another hero in the study of electricity, which showed that electricity passes instantaneously over a long length of wire. Franklin, he said, had passed a current through many miles of wire, but had observed no passage of time between the closure of the circuit at one end and the spark obtained at the other.

In such case, observed Morse, he could see no reason why intelligence might not be instantaneously transmitted by electricity to any distance.

It's important to note that although Morse thought he had come up with a new idea, this was certainly not the case. The basic idea of the telegraph, and even some actual devices, had already appeared on the scene, though Morse apparently was blissfully unaware of this. But this misconception may have been the best thing that could have happened to him, and to the world at large. If he had known that his was not a new idea, he might never have become obsessed with the telegraph. As it was, he spent the rest of the long voyage deeply involved in developing his initial ideas. He put these down in a notebook, which came in handy later on during the many disputes to follow.

By the end of the six-week voyage, Morse had a design for a sending-and-receiving apparatus and even a code for converting simple signals to numbers and then to words. Both the design and the code were to be strongly modified before the final designs came into being. But both provided solid foundations for later work.

## *Contesting Priority*

Competition for priority emerged early. Only a few weeks after the *Sully* docked, Dr. Jackson wrote to Professor Silliman. "On my voyage home," he stated, "I had the pleasure of becoming acquainted with S. F. B. Morse, a distinguished American artist, who is very ingenious in mechanical inventions. We employed our weary hours at sea in contriving various things; among which, *we invented* an Electric Telegraph."[2] (Italics mine.)

There his partial claim lay until, some five years later, as word of Morse's telegraph began to spread, Jackson upped it to a claim of being the *principal* inventor, and finally, to being "*the* inventor." Here's a brief summary of the progression.[3]

In a letter from Jackson to Morse, dated November 7, 1837, Jackson wrote, "On my return from the forests of Maine last Friday I found your letter of 18th September last, which contains a claim to the Electro-Magnetic telegraph as your own exclusive invention. This claim of yours is to me a matter of surprise and regret, for I have always entertained the highest opinion of your honor and fairness, and should be very sorry to have any reason to change my opinion of your character. It becomes me, however, to claim and to sustain that portion of the honor of the discovery and invention which is my due."

He says later in the same letter, "You will acknowledge that you were at that time [on board the *Sully*] wholly unacquainted with the history and management of electricity and electro-magnetism, while I was perfectly familiar with the subject, it having been one of my favorite studies from boyhood to the present hour, and I had enjoyed every possible advantage in acquiring a full knowledge of the subject during my studies in the scientific schools of Paris and elsewhere."[4]

Finally: "Hence, since I had performed all the experiments in detail, and had here brought [them] together for a specific purpose . . . I do claim to be the principal in the whole invention made on board the Sully. It arose wholly from my materials, and was put together at your request, by me. This you certainly will not pretend to dispute."[5]

Well, of course Morse did dispute this, and hotly. "Your memory and mine," he answered, "are at variance in regard to the first suggestion of conveying intelligence by electricity. I claim to be the one who made it, and in the way which I stated in my letter to you."[6]

"Now, sir," Morse continues, "I not only deny that all the materials were furnished by you, but I deny that I am indebted to you for any single hint of any kind whatever which I have used in my invention. . . .[7] I have chosen to ascribe your motives, in asserting so unfounded a claim to my labors, to honest forgetfulness of the circumstances as they occurred on board the *Sully,* and to a misconception of the nature of my invention. And I trust that no measures on your part will compel me to adopt any other less favorable explanation of your conduct. I therefore yet hope that this matter can be settled in private."[8]

Well, this was not to be, either. Not long after, when Morse was in Europe pursuing his hoped-for rights on that continent, Jackson engineered an article in the *Boston Morning Post*. I say engineered, because the article read like an editorial comment:

> We are informed that the invention of the Electro-Magnetic Telegraph, which has been claimed by Mr. S. F. B. Morse, of New York, is entirely due to our fellow-citizen, Dr. Charles T. Jackson, who first conceived the idea of such an instrument during his return voyage from Europe in the packet ship Sully, in October, 1832.
>
> Mr. Morse being his fellow-passenger, and having pretended to feel a great interest in the invention, and a desire to participate in the experimental trials of the machinery, Dr. Jackson freely communicated to him and to all the cabin passengers his various plans for effecting the telegraphic communications.
>
> Subsequently, Mr. Morse undertook to monopolize the credit of the invention, when Dr. Jackson wrote him a friendly remonstrance, which [resulted in] a long sophistical reply of a very unsatisfactory character. This was followed by a severe reprimand from Dr. Jackson with a detail of all the circumstances of the invention, and of the conversation which took place on board the Sully, when an impudent answer was returned, giving Dr. Jackson to understand that Mr. Morse had taken out a patent for the invention, and was the only inventor known to the laws, and cautioning him as to his proceedings. . . . The origin of the idea of the new Telegraph, as above stated, can be proved by a number of the passengers on board the Sully at the time.[9]

Later, in the midst of other exchanges as well as in several patent trials, Morse's opponents did try this route, but somehow the *Sully* passengers never seemed to remember things in the same way.

As if the *Post* article were not bad enough, Jackson also wrote to a member of the French Academy of Science and made similar claims.[10]

Suffice it to say that Jackson, brilliant but extremely erratic, later got involved in other violent controversies when he claimed priority in such matters as the use of ether in anesthesia and in the invention of guncotton, an explosive that found wide use as a smokeless powder. He finally suffered a complete mental breakdown and died seven years later in a mental hospital.[11]

Unfortunately, Jackson's various claims were nevertheless appealed to several times later on during other battles and although they did not seem to have much effect, were the cause of considerable aggravation to Morse.

## *Morse's Design*

Morse's initial design was workable, but barely, and was considerably improved as time went on. What distinguished it, at least at first, from its predecessors and competitors was that he envisioned a system that would record messages at the receiving end.

Competitive devices already under development—and there were several—mostly depended on one of the new electrical discoveries, namely that a compass needle placed next to an electrified wire will move when the current is turned on and off. But the resulting devices, called needle telegraphs, had to be constantly monitored.

Morse's first concept for a code involved a simple set of signals, plus spaces, for just the ten common digits, 0–9. These were what would be transmitted. Combinations of these digits would represent words from a list contained in a special dictionary. So before a message could be sent, it first would have to be translated into numbers.

To send the signals for these numbers, Morse envisioned preparing for each message a long bar or stick that had teeth inserted along the bottom edge, with the teeth representing the needed digits. As the bar was passed through the sending apparatus, the teeth would cause a lever to rapidly open and close the electric circuit.

At the receiving end, a pivoted bar was actuated at one end by an electromagnet that was controlled by the sending apparatus. At the other end of the bar, a pencil recorded the pulses of electricity on a paper tape that was pulled along at a constant rate. A brief contact produced a single sawtooth break in the resulting penciled line; two consecutive contacts produced a double sawtooth; a short pause produced a space between numbers that were to be combined (e.g., 2-1-5); and a still longer pause produced a space between adjacent single numbers or combinations.

The resulting sawtooth marks could be retranslated back to numbers and then to words via the same dictionary. It was somewhat cumbersome, but still better than what had come before. A competing system in Europe, which had actually been described to Morse by Jackson, involved a periodic, brief chemical reaction at the receiving end. This method found some use for a while, until updated Morse designs showed their true potential.

Arriving home, Morse, fired up by what he took to be his new invention, could envision it spreading across the world. He knew that no single, unbroken wire could possibly transmit an electrical current over such a distance. But Jackson had also been able to describe to him the work done by none other than Michael Faraday (see chapter 2). Faraday had shown that electric current in one circuit could generate an electric current in another circuit (electromagnetic induction). This formed a basis for the relay concept—that it was possible for a signal that had weakened during its passage to be strengthened and passed on to the next stage in long-distance communication and thereby to potentially span the world. It was a breathtaking idea.

At the same time, Morse also saw the telegraph as a way of funding his art career, which he still felt was his major calling. Torn between the excitement of his new idea and the realities of earning a living, he tried to do both when he returned to the United States. But he had neither the resources nor the skills to bring his invention to the public, and spent several frustrating years trying to be both artist and inventor, as well as family man. He had married in 1818, and though his wife died only seven years later, he did have three children, who often had to be cared for by others as Morse struggled to balance his life and career. (It was not until the summer of 1847 that Morse would finally have his own house, with his three children with him.)

In 1835, he was appointed professor of the Literature of the Arts of Design at New York City University (later to become New York University). Nevertheless, the next year was one of the darkest in his long, troubled life. Teaching art while thinking all the while about his invention, he was tormented by his inability to move his project along. Too, he was spending so much of his meager income on materials that little was left for food and other essentials.

On the other hand, he now had decent quarters and finally had some space to work in. His studio, again, became a laboratory. He immediately began work on a prototype of his telegraph, and did indeed come up with one—built around an old picture frame. It was an astonishing accomplishment, produced after much sweat and expense; but it had serious limitations, including the fact that it worked only for very short distances, on the order of tens of feet. He had no idea what the problem was.

## *Other Inputs*

Had Morse been more conversant with the ongoing activities in the world of electrical science, he would have been aware that one of his own countrymen, the physicist Joseph Henry, had solved Morse's problems some five or six years earlier. Henry had constructed a powerful electromagnet and, setting up a long circuit with the electromagnet at one end, had rung a bell at the other end. In writing up his work for publication, he had even suggested its possible use in telegraphic work. He had not, however, come up with a full design; for Henry was a scientist, not an inventor, and had little interest in continuing to work along these lines. Once he had done his experiments, he went on to other matters, even though his colleagues had urged him to patent his apparatus and the various applications it made possible.

Then, on an afternoon in January 1837, Leonard D. Gale, a chemistry professor at Morse's university, saw Morse's primitive apparatus, and, fascinated, began to supply the scientific background that Morse sorely needed. He pointed out that Morse's battery was too weak by far. But a battery then was something very different from a battery today. Called a pile or cup, it was made up essentially of alternating copper and zinc disks lying in an open bath of salt or, later, acid. Each pair of disks, separated by nonconducting material such as cardboard, constituted a single cup or cell. Morse was using a single large cup. Gale said right away, "You need more cups." He suggested using a battery consisting of twenty cups.

He also pointed out that Morse's magnet was not up to the task; and when Morse added hundreds more turns of wire around the iron core of his electromagnet, the device began to show its mettle. Morse, recognizing his limitations, invited Gale to join him in the enterprise. Gale became an official partner and received a one-quarter share in the patent and the profits.

The partners decided it was time for a public demonstation. For that, however, they needed a much better prototype. But neither man was up to the task. Again, someone came to the rescue. Alfred Vail, who was a friend of Morse's, had seen the work developing. He also had been working at his father's foundry, located in Morristown, New Jersey. With the approval of his father and brother, Vail offered the facilities of the firm, plus the financial wherewithal to make it all happen. Vail was cut in for another one-quarter share of the enterprise, plus half of all income from activities that should materialize in France, England, Scotland, and Wales. All improvements and patents were, however, to be honored with Morse's name.

Things began to move more quickly. With Vail's help, the group produced a much-improved prototype. It came time to try to obtain patent protection for the new device. Morse filed the first application in the fall of 1837—under his name, of course. Shortly thereafter, the partners came up with a much simpler and more direct alphabet code, of the type we know today. The new code eliminated the need for the conversion dictionary and was a huge advance in the practicability of the system.

The next step, they felt, was to petition Congress for financial aid in constructing a full-scale public demonstration. But while the proposal had some supporters, it was mocked by many.

Congress finally came through, but it took five long years. During this time, the partners, all but giving up on government help, tried to raise private capital. But what businessman in his right mind is going to invest in such a weird idea?

As Joseph Henry's colleagues began hearing about Morse's application to the government, however, they urged Henry to object that he had put forth the idea before Morse. In fact, they felt that although Henry had not applied for a patent, his work had put the idea into the public domain. Henry actually considered doing this, but an accidental meeting with Morse changed his mind. Henry became convinced that Morse, who already had put so much into this new device, deserved to win the government's backing, and he offered no objections.

At a later meeting, requested by Morse, Henry offered help and information on the next hurdle, which had to do with just how far each segment of a telegraph line could reach before the information became weak enough to lose its accuracy. On board the *Sully,* Jackson had told Morse about Faraday's work with electric current, but as usual, getting something to work is harder than talking about it; and Henry helped out here as well.

Careful experiments had shown that a single circuit could work at distances of hundreds of miles. But for a coast-to-coast, or transatlantic, line, this would not be good enough. Henry gave Morse additional information that led, eventually, to what today we would call a relay.

Faraday had made a basic discovery in 1831, which is what Jackson apparently told Morse about on board the *Sully.* Henry's work with electromagnetic induction, which brought him international fame, carried this process further. A true relay would take a weak telegraphic signal and automatically rejuvenate it to the point where it could be sent on its way with renewed vigor. This might have to be done over and over again when setting up a long line. Henry's relay, based on electromagnetic induction, was another step along the way but was not well suited for this purpose. In his laboratory experiment he had used a secondary circuit to break a primary circuit that was powering an electromagnet holding up a heavy weight. But the primary circuit would not reset automatically. He had to do that by hand. In other words, it could not cycle. Once again, we see a case in which Morse took some basic developments and created a working apparatus—what turned out to be the first actual relay.[12]

In the meantime, Morse learned some disquieting news. Telegraphic activities in Europe were heating up as well. Henry had already seen and appraised a system that had been developed by Charles Wheatstone. Like a few others that had already come along, it was based on a pivoting needle. This made it very sensitive, but as Henry was later to point out, it was not nearly as practical as Morse's system.

By 1842, the Morse team made a final try at gaining government support, though their chances still looked rather bleak. Morse was advised to obtain letters of support from recognized scientists. He once again appealed to Henry, who answered with the requested statement of support on February 2: "Science," he wrote, "is now fully ripe for this application, and I have not the least doubt, if proper means be afforded, of the perfect success of the invention."[13]

So it would appear that Henry was unstintingly supportive to Morse's struggle. But, says the Henry biographer Albert E. Moyer, there's more to the story. He points to Henry's somewhat grudging tone in the same letter, which "relegated Morse's telegraph to the subordinate status of a mere practical application dependent on more critical, fundamental scientific principles."[14] Morse, happy with the endorsement for the telegraph, was willing to ignore, for the time being, Henry's description of the invention as a rather derivative development.

Henry's endorsement of the telegraph as a new device may well have made the difference. The government subsidy did come through: $30,000 for a demonstration line between Baltimore and Washington, D.C. Elated, Morse's group began to lay the line—underground, but based on a misconception that guaranteed its failure. After considerable funds were spent, tests showed that there was a serious problem with the insulation on the telegraph wire. Perhaps the tubing leaked and let in moisture; or maybe the insulation was bad. In any case, Vail suggested using overhead lines, but his idea for insulating the line from the poles was also faulty. A suggestion from the works manager, Ezra Cornell, was ignored—until Morse once again consulted Henry and got some more good advice. Henry pointed out that the problem had already been faced and answered elsewhere with use of simple glass insulators—a method that remained in use for decades thereafter.

After much struggle, the line was built and successfully demonstrated in May 1844. Telegraphy was no longer a crackpot idea.

## *A Serious Rift*

Again, Henry had come to the rescue. The relations between Morse and Henry seemed cordial enough. But nothing lasts forever, and unhappily, a series of events took place that caused a serious rift between them. It began when Vail, who had been in on the development almost from the beginning, decided to write a history of the electromagnetic telegraph in the United States. It was published in 1845 under the title *American Electro Magnetic Telegraph.*

Henry saw it and quickly noted that there was no mention of the part he had played in the development of what was clearly becoming a major new invention. Yet it did include the supporting letter that Henry had written for Morse. Henry assumed that Morse had to have had a part in the book's preparation, and did not hesitate to let his displeasure be known. He told his classes, for example, that he, not Morse, was the inventor.

Morse, however, had been in Europe, trying to cement his claim there, during Vail's preparation of the book. Henry continued to complain. When Morse returned to the United States and learned of Henry's unhappiness, he tried to determine what had gone wrong. Vail told him that he had had difficulty finding anything in print about Henry's contributions and had tried, unsuccessfully, to obtain an account of Henry's work via one of Henry's students.

Another point might explain Vail's actions: Before he left for Europe, Morse had suggested that Vail not give away too much in describing their telegraph's receiving magnet. He had not, however, intended that Vail should omit Henry's basic work on electromagnets.

Vail tried to exonerate Morse and requested that Henry point out such omissions so that they might be rectified in the next edition. Henry was so angry that he never even replied to this request, assuming, correctly, that Morse could supply the information. But in spite of all the excuses and explanations, Henry refused to believe that Morse really had no knowledge of what was going on.

By this time, many of the early kinks in the Morse system had been ironed out, and the group's company, now called the Magnetic Telegraph Company, was expanding rapidly. Morse's vision of a giant net was well on its way to being realized. By 1846, telegraph lines ran from Portland, Maine, to Washington, D.C., and west to Kentucky and even Wisconsin. Within another two years, every state east of the Mississippi with the sole exception of Florida was linked by telegraph.

## *Patent Battles*

But, as often happens when a new and exciting business opportunity presents itself, eager entrepeneurs began trying to profit from the new invention, often without paying the licensing fees that Morse and company were demanding. In several cases, lines had actually been put up and equipment installed, particularly in some out-of-the-way regions of the country.

As a result, Morse went to court several times. The defendants, who at first had tried to elude his patent, now tried to break it. Their basic defense was that Morse was not the inventor and that the telegraph should be in the public domain. They pointed out, for example, that with the advent of the alphabet code in 1838, telegraphers found they could decipher the clicking sounds directly and write out the message as it came in. This not only speeded up the process considerably, but also eliminated the need for Morse's recording apparatus (which he at first felt distinguished his system from others but which his system was no longer using, either). Morse did include the acoustic method in a subsequent application.

The defendants also called on Henry to testify in no less than three separate cases. Always reluctant to get involved in such affairs, he had no interest in going after Morse. But when called on to testify, Henry essentially stated in all three cases that much of what Morse had used had been known or created earlier by others. At the same time, he still felt compelled to praise Morse for having come up with a practical system, and one better than anything that had been built before.

Henry's equivocal stance perhaps can be better understood in the light of a point made by his biographer Moyer: "[I]n public forums or professional exchanges with all but his closest inimates, Henry always maintained a stance of modesty and propriety."[15] In private, as we saw earlier, he could complain, and loudly.

All the cases were decided in Morse's favor. But telegraphy had become important enough by mid-century that one of the cases was fought all the way up to the United States Supreme Court. The decision, which came in 1854, was again in Morse's favor.

Morse's son, who edited Morse's letters and journals, later asserted not only that Henry's testimony was offered reluctantly, but that it was tinged with bitterness, mainly caused by Vail's treatment of him in his book, and also because of what he considered to be Morse's exaggerated ideas of what he had accomplished. Henry, it should be noted, had been paid for some consulting work by one of these upstart companies.[16]

## *The Fruits of Success*

By now, Morse was tasting the fruits of success—admiration, flattery, and wealth—and was loving it. In addition, he had moved far along in convincing himself that he alone deserved full credit for the invention. In an important early biography of Morse, Samuel Prime wrote, "He believed himself an instrument employed by Heaven to achieve a great result, and, having accomplished it, he claimed simply to be the original and only instrument by which that result had been reached."[17]

Oddly (to my mind), Prime saw this as a positive characteristic, for he added, "With the same steadinesss of purpose, tenacity, and perseverance, with which he had pursued the idea by which he was inspired in 1832, he adhered to his claim to the paternity of that idea, and to the merit of bringing it to a successful issue. Denied, he asserted it; assailed, he defended it."[18]

On the other hand, Henry's halfhearted, grudging testimony, which included statements regarding others' priority in several areas, sat less and less well with Morse. Further, Henry's growing reputation meant that his comments were being heard ever louder and clearer in both the United States and Europe. This was worrisome to Morse, who was still facing plenty of opposition.

Further, he had applied for an extension of his 1840 patent, which would include some of the group's new developments. Regardless of his feelings about Henry, he felt he couldn't very well attack this respected gentleman in public while his patent extension was being considered. So at first he was rather circumspect. When he described Henry's work as "jackdaw dreams," it was in private, via a letter to his colleague Alfred Vail.[19]

After the extension was granted (1854), and after the court cases were settled, Morse felt easier about speaking his mind, and figured it was time to cement his claim to be the sole inventor. So in 1855, there appeared, in the journal *Telegraph Companion,* his ninety-six-page testimonial to himself. It was titled "Defence against the injurious deductions drawn from the deposition of Professor Henry."

One of Morse's comments gives the article's flavor: "I shall show that I am not indebted to him [Henry] for any discovery in science bearing upon the telegraph; and that all the discoveries of principles having this bearing, were made, not by Professor Henry, but by others, and prior to any experiments of Professor Henry in the science of electro-magnetism."[20]

Henry was furious. He wrote to his brother-in-law, referring to Morse as "the telegrapher," and complained of his "libelous pamphlet [which was intended] to put me down in the eyes of the public. Morse is a fool in this, as well as something worse; having seen by the papers that I was attacked [in an unrelated matter, having to do with the Smithsonian Institution] and supposing that I would be weakened, concluded, coward as he is, to jump on me."[21]

He also maintained elsewhere, "I am not aware that Mr. Morse ever made a single original discovery in electricity, magnetism, or electromagnetism, applicable to the invention of the telegraph."[22]

## *Thrust and Counterthrust*

We know that Morse was heavily indebted to Gale, Vail, and Henry; yet as time went on he gave out less and less credit to them. But it was Henry at this point who felt he had the most to lose. As a shining light in the newly developing world of American science, he felt that any blot on his reputation was unbearable. Perhaps he was annoyed with himself for not going to the trouble of patenting his work, and wanted at least the credit.

But, perhaps most important, he was no longer an academic who could afford to swallow Morse's charges even if he did not wish not to get personally involved. By now, in recognition of his growing eminence, he was the head of the newly forming Smithsonian Institution. He felt that Morse's accusations surely would interfere with his effectiveness in that office. This he could not permit. How best to proceed?

Rather than go to the public, as Morse had, or publish as an individual in a scientific journal, Henry appealed to the Smithsonian's Board of Regents and asked that a committee be appointed to investigate the charges. He not only gathered material on his own, but supplemented it by testimony from other witnesses to the history of the device.

Among those he appealed to was Professor Gale. Although Gale had stated in earlier testimony that the Morse design in 1836 was essentially the same as the practical device that the world came to know, he had under challenge changed his mind. He now stated that Morse's device had indeed needed some serious help.

Thanks to this testimony, plus much more, the committee came through with all that Henry had hoped for. Their report declared that Morse's "Defence" was actually "an assault upon Professor Henry; an attempt to disparage his character; to deprive him of his honors as a scientific discoverer; to impeach his credibility as a witness and his integrity as a man. It is a disingenuous piece of sophisticated argument, such as an unscrupulous advocate might employ to pervert the truth, misrepresent the facts, and misinterpret the language in which the facts belonging to the other side of the case are stated."[23]

Henry was thereby exonerated much more effectively than if he had tried to do this by publishing on his own. He was then able to let the matter go and to apply himself to the major business of running the institution.

Morse now felt that he was the aggrieved party; but he also let the matter go—for a while. Ten years later, however, in writing a report on the electrical section of the Paris Exposition, he tried to repeat Vail's trick of making Henry disappear from the history of the telegraph.

But the editor of the reports, W. P. Blake, would not accept this and strongly urged Morse to remedy this omission. Blake even thought he might engineer a reconciliation. Morse, however, would go no further than a grudging addition: "In more recent papers, first published in 1857, it appears that Professor Henry demonstrated before his pupils the practicability of ringing a bell by means of electro-magnetism at a distance." Blake gave up on the reconciliation.[24]

When Morse died in 1872, many important people attended his funeral. Henry was not among them.

The whole affair was particularly sad in light of their initial good relations. The Smithsonian Regents committee made a good point in their summation:

> It appears that Mr. Morse was involved in a number of lawsuits, growing out of contested claims to the right of using electricity for telegraphic purposes. . . . The testimony of Mr. Henry, while supporting the claims of Mr. Morse as the inventor of an admirable invention, denied to him the additional merit of being a discoverer of new facts or laws of nature, and to this extent [Henry's testimony] perhaps was considered unfavorable to some part of the claim of Mr. Morse to an *exclusive* right to employ the electro-magnet for telegraphic purposes.[25]

This may well have been the case. And, considering what Morse was going through in his long and continuing state of siege, his position, which seems unreasonable at first, may make some sense after all. After struggling for over a dozen years to bring the invention to fruition, he then faced one battle after another for another six.

Among the witnesses called in was Dr. Charles Jackson, who was happy to reiterate his claim that he, not Morse, was the true inventor. The courts listened to Jackson's claim, but fortunately, gave it no weight.

The courts, nevertheless, did find themselves struggling with the question of whether Morse was indeed the inventor of the electromagnetic telegraph and whether he was entitled to license fees from his patent. Henry's testimony, which distinguished between the invention and the principles behind it, was central to their decisions. The likely reasononing was later stated nicely by Leonard Gale, who remained in Morse's corner: "It has been said," he commented, "that Morse could never have made his invention had he not availed himself of what Henry had done. This argument amounts to nothing, because every inventor uses the facts of science known to the world."[26]

Before the patent battles, Morse seemed to have no hesitation in acknowledging Henry's contributions. He may later have feared, perhaps rightly, that such admissions could be damaging to his chances. Further, the press and the public, hearing mainly the objections in the court cases (and perhaps with a common American tendency to side with the underdog), tended to look at Morse as greedy, and generally sided with his opponents.

Once he had started to believe that he was indeed the sole and true inventor, it apparently became easier and easier for Morse to maintain the idea, until he became quite convinced of its truth. While he had earlier been quite willing to credit Henry for his considerable help, he was now denying it entirely.

## *Origins of the Idea*

Clearly, some of Morse's right to priority has to do with *when* he actually began thinking along these lines. He believed, and continued to maintain, that it all began on the *Sully.* Ironically, others argued that he actually had the idea before then. A friend of Morse's, R. W. Habersham of August, Georgia, had spent some time with Morse in Paris early in 1832. He later maintained that even before the *Sully* voyage, Morse, totally disgusted with the slow pace of ordinary mail from the United States, already was turning over thoughts of how to speed up the process. Habersham remembered Morse pointing out that the French system would do better in the clearer atmosphere of the United States, but that it was still too slow, and that lightning (a common name for electricity in that day) could do the job nicely. Morse, oddly, never was able to recall this conversation, though it surely would have helped him in his later battles.

Prime points to another, curious, bit of evidence supporting Habersham's claim. Morse and the American writer James Fenimore Cooper had spent time together in Paris before the *Sully* voyage. Although Cooper's novel, *The Sea Lions*, was not published until 1849, it dates an early discussion of the telegraph to the winter of 1831–1832. The narrator states, "[W]ell do we remember the earnestness, and single-minded devotion to a laudable purpose, with which our worthy friend [Morse] first communicated to us his ideas on the subject of using the electric spark by way of a telegraph. It was in Paris, and during the winter of 1831–1832, and the succeeding spring, a time when we were daily together, and we have a satisfaction in recording this date, that others may prove better claims if they can."[27]

While Morse's legal battles ended in 1854, the arguments pro and con went on for years, and brought in other combatants. F. O. J. Smith, who initially had been an important member of Morse's team, changed sides and came up with a lovely, if somewhat questionable, metaphor: "While Henry was strictly the father," Smith opined,

> Gale was no less truly the midwife, at the birth of the American Electro-magnetic Telegraph. Professor Morse, in fact, only acted the part of the errand boy, who called in the midwife's service, to save the life of the unborn child; and even after its birth, it was too feeble and slow of motion, too deformed in *limbs* and *speech,* to be of value, without the nursing and ingenious new mechanical appliances of Mr. Vail, or some equivalent artisan, and it is sure that without them it could never have grown into manhood, or have been utilized for business purposes.[28]

It's nicely put, but I'm not convinced of its correctness. Smith and others also argue for the primacy of Henry's claim, based not only on his basic work, but on his mention in his 1831 paper of possible telegraph use. Although his refusal to apply for a patent makes him sound very high-minded and altruistic, one of his biographers (Moyer) presents an additional reason. Henry, he argues, really felt that these discoveries and devices were "philosophical toys," not likely to find use in the tech world, for new developments in coal-fired engines seemed to him to be far more efficient and less costly. Morse had a much better fix on the possibilities.

## *Morse's Contributions*

Here, perhaps, is the place to stop and really try to evaluate Morse's contributions. There is no question that Henry, Gale, and Vail played important parts. Gale and Vail both profited financially from their association with Morse, whereas Henry certainly was rewarded in other ways for his broad contributions to the world of electrical science.

And Morse. What exactly did he contribute? In my order of importance:

- A rare combination of artistic creativity and technical ingenuity, which enabled him to come up with the first practical electro*magnetic* telegraph. Other telegraph systems that preceded his were in essence electric telegraphs, inferior in both concept and practicality.

In addition, he must be credited with:

- A vision, early, of the promise and potential of such an invention, as opposed to Henry's concept of a philosophical toy.
- An ability to accept his limitations and to offer rewards to others, when necessary, for their help.
- An ability to find the right people to fill in for the holes in his technical armamentarium.
- Personal and political connections, needed both to overcome continual opposition and competition, and to gather the necessary financial support.
- An organizing ability, that is, the ability to bring together a variety of disparate elements and fashion a finished product.
- Coming up with the best alphabet code yet devised in 1838. Although some say the improved code was Vail's, others argue that Morse was working on an alphabet code earlier than 1838.[29] Morse's 1844 code eventually became the world standard.
- Finally, a dogged resolution that kept him focused fiercely enough and long enough to make the whole thing happen.

## *Fallout from the Dispute*

The festering dispute between Morse and Henry may have had one very positive outcome. As I stated in the last chapter, the United States in the early nineteenth century was a "hands-on" kind of place, and it remained so for a considerable part of the century. Land was cheap and skilled labor expensive, leading to a strong incentive to develop labor-saving devices.

New inventions flowed quickly from the practical minds of its inhabitants—a remarkable cornucopia ranging from Eli Whitney's cotton gin to McCormick's reaper, and on up to Edison's incredible outpouring of inventions around the turn of the twentieth century.

The problem, in Henry's opinion, was that our eye was on the product rather than on the science behind it. We still lagged far behind Europe in the world of basic research. It was this distinction that he tried to make in his several statements when called upon to testify in the court battles. And it was to rectify this unbalance that he now turned his attention as head of the Smithsonian.

The biographer Moyer writes:

> His [Henry's] grudge against the two telegraph entrepreneurs [Morse and Vail] undoubtedly gave immediacy to his call for emphasizing basic research. Thus, after underscoring the nation's neglect of research, he shifted to the converse of the problem—the nation's disproportionate and misguided

> attention to practical applications. Henry likely had in mind popular accounts of the telegraph—particularly Vail's recent book—that glorified the technical invention while disregarding the underlying science. "Indeed," he [Henry] complained, "original discoveries are far less esteemed among us than their applications to practical purposes, although it must be apparent on the slightest reflection that the discovery of a new truth is much more difficult and important than any one of its applications taking singly." Smarting under the perceived slight from Morse and Vail, Henry wanted the Smithsonian to redress the national misplacement of values by bolstering support for basic research.[30]

Henry assumed the secretaryship in 1846. Not long after, the *New York Observer* ran an article on the new institute and its new director. It included an encomium by an unnamed person—who turned out to be none other than Morse himself, apparently still trying to make amends. No one knows for sure whether Henry ever actually saw the *Observer* article, but the two did meet not long after, and although Henry was still angry about Vail, he and Morse did manage a reconciliation of sorts. Henry was left with the impression that the next edition of Vail's book would include satisfactory coverage of his contributions.

Unhappily, only a few months later, another printing of the book was released. Though the date was changed, it was in reality the same book, from the same plates. Nothing else was changed. As far as Vail and Morse were concerned, it was simply an expeditious route to putting out what seemed to be a new edition.

Henry saw the new date, but not anything on himself, and decided enough was enough. The break was now complete and permanent. But Henry, now in a position of great prominence, got his revenge on Vail. The cornerstone for a major Smithsonian building was to have included Vail's book. Henry convinced the organizing committee to eliminate it from the collection of historical items to be buried in the stone's hollow core.

Henry served the Smithsonian honorably and well until his death in 1878. A sixteen-foot statue of him was erected in 1883 and still stands today at the entrance of the Smithsonian Institution. It is a fitting companion to, albeit a comfortable distance away from, Morse's memorial statue in New York's Central Park, which had been erected a dozen years earlier.

# CHAPTER 4

# *Edison versus Westinghouse*

## The AC/DC War

For the better part of recorded history, campfires, torches, oil lamps, and candles were the only usable sources of night lighting. The introduction of gas lamps provided a major advance, and by the mid-1800s, cities all over the world were lit, if they were lit at all, by gas flames, with the gas distributed by underground pipes. As evening approached, the lamps would be lit one by one, and a new buffer was put up against the fearful dark. The gas, manufactured from coal, was also piped into homes and factories. A major industry had grown and prospered.

Yet manufactured gas, like firedamp (natural coal gas), was not only flammable but explosive; and accidents were, as in the mines, a major threat.

Half a century earlier, however, the electric battery had been invented, and experiments showed that certain materials could be made to produce heat by passing electricity through them and, if made hot enough, light. Humphry Davy began a small revolution in the field of lighting when he drove electricity through a pair of carbon electrodes separated by a small gap. The result—a kind of continuous spark in the gap—was called arc lighting, and it found some use in place of gas lamps. But the light was uneven, and the electrodes burned out too quickly and had to be constantly replaced.

The result was that manufactured gas had become the major player in night lighting. At the same time, railroads, steamships, and many factories still relied heavily on coal-driven steam engines for motive power. There were some people, however, who could see further ahead. George Stephenson, for example, predicted early in the century that electricity—still basically an experimental toy—would one day become the main source of power all over the world.[1]

He was right, of course. What he couldn't foresee was that the early days of electricity's development as a world force would see a battle that surpassed even his own with Sir Humphry Davy earlier in the century. But whereas Davy and Stephenson were both working on improving the same basic lamp, the "war of the currents" was fought over very different technologies. These were Thomas Alva Edison's entry, direct current, and that of George Westinghouse, alternating current.[2] These terms are commonly written as DC and AC.

It was a war that was fought on several fronts—between the two men, who were both highly competitive; between the companies they founded, leading to an astounding array of patent battles; and between two very different technologies, both aiming to accomplish the same task—electrifying the world. It lasted from the time Westinghouse chose to challenge Edison in New York City in the late 1880s to the cross-licensing agreements arranged not by them but by their business associates, a decade or so later.

## *Edison*

Thomas A. Edison was born in Milan, Ohio, on February 11, 1847. Though Stephenson and Edison came from different generations, there was considerable similarity between the two men. Like Stephenson, Edison had little formal education, but was ambitious, largely self-taught, technically adept, a tinkerer from the word go, and wonderfully enterprising. By age twelve, he had a regular job selling snacks and newspapers on the Port Huron/Detroit branch of the Grand Trunk Railroad.

A bright and inquisitive youngster, he ferreted out all kinds of information during his travels and layovers. With the purchase of a small hand printing press, he began to write, publish, and print his own weekly newspaper, which featured both local gossip and news items about the Grand Trunk Railroad and its service. Although the paper would hardly win any prizes for grammar and spelling, he nevertheless managed to build up a subscription list of several hundred customers among the passengers and employees of the railroad.[3]

As if that weren't enough, he also convinced the train conductor to grant him permission to carry out chemical experiments on the train during his spare moments.

During long layovers in Detroit, his favorite hangout was the railroad telegraph office, where he picked up the rudiments of telegraph operation. Then, in the fall of 1862, he saved the life of the young son of James Mackenzie, the station agent at one of the train's stops. As a reward, Mackenzie gave him regular lessons in telegraph operation. A couple of months later, at the age of sixteen, Edison had a job as an operator in a Port Huron office.

But Edison was hardly the type to be satisfied with routine office work. Thanks to his inclinations to tinker with the equipment, plus his love of reading (apparently, often when he should have been working), he lost this and several subsequent positions and became what was called a tramp operator, meaning he filled in here and there as needed.

Operators at that time were required to maintain their own equipment, which turned out to be a blessing for Edison. Like Samuel Morse, he filled notebooks with drawings of his own designs arising from his wide reading (including Faraday's research on electricity) and his equally wide experience with equipment.

Then, in 1864, at the age of seventeen, he came up with a design for an automatic telegraph repeater, which made it possible for the system to receive and store messages at high speed, then repeat them at a slow enough pace for the operator to decode comfortably. It was the first of a long series of inventions and improvements that were to come from his efforts. Many of his early inventions had to do with the telegraph, including a stock ticker and a printing telegraph.

On January 30, 1869, just nine days after he sold the rights to the printing telegraph, he placed a small notice in the *Telegrapher,* a trade journal, announcing that Thomas A. Edison, formerly a telegraph operator, "will devote his time to bringing out his inventions." This he did, and over the next sixty-two years, he amassed the incredible total of 1,093 patents granted.

Many of his later inventions also came out of his work in telegraphy. The phonograph, one of his most famous inventions, grew out of his work with the earlier automatic telegraph repeater. Just how closely did Edison identify with telegraphy? He nicknamed his first two children Dot and Dash.

Like Stephenson, he saw himself as a practical man and even made fun of education. In his later years, he claimed, with only a slight exaggeration: "School? I've never been to school a day in my life. D'you think I would have amounted to anything if I had?"[4]

He did indeed amount to something. By age thirty, he already had established several successful businesses for manufacturing and selling his new inventions, and had a working research laboratory with able assistants to help in his inventing. He had, however, little interest in the basic science of any development, or in what might come out of it years later. He was interested in the needs of the present. One of these needs was for a better method of lighting up the night than the still-primitive method of gas lamps that, in spite of their many drawbacks, remained the main method of illumination in the cities' streets and buildings.

Once he sank his teeth into a problem, he wouldn't let go. The story about how he tried hundreds of different materials in his feverish search for a usable, practical electric light, and his eventual success in 1879 with a carbonized filament, has been told many times. Not always told is the fact that others had worked in this area, and even had come up with working incandescent bulbs. For example, Joseph Swan, a British inventor, exhibited such a lamp in 1875. Swan's lamp was a good start, but it had a variety of disadvantages, including the fact that it burned out quickly. Nevertheless, after Edison developed his own bulb, he set up a partnership with Swan in 1883 for production and sale of their bulbs in England.

But Edison could see a bigger picture. Lighting a single lamp in a laboratory, or even lighting up a string of lamps, was a good start. But properly distributing electric current for great numbers of electric lamps, of different strengths and sizes, that would constantly be turned on and off, was quite another matter. To make any lamp feasible, he not only would have to improve on the Swan design, he would have to create a whole new industry. That is, he would have to develop a constant-voltage dynamo (which, combined with a steam engine, converted mechanical energy into electricity), plus an array of other equipment, including insulated conduits, main lines, underground junction boxes, relay circuits, and switchboards. At the user's end, he needed new meters, fuses, fuse boxes, and even sockets for the lamps. For the manufacturing process, he had to find a company that could make the fragile glass bulbs and set up a process for setting the filament and its connections in a vacuum within the bulb.

He did all of this. It was an amazing accomplishment, and it culminated in several public demonstrations that knocked the socks off everyone who saw them. First came a small demonstration at his Menlo Park, New Jersey, laboratory on December 31, 1879. Then another on the steamship *Columbia,* which, Edison figured, would spread the word during its travels from city to city.

The real excitement, however, was generated by his famous Pearl Street Station in downtown New York. On September 4, 1882, he flicked a switch, and a ten-block area suddenly blazed into light. In the Menlo Park demonstration, the wires had been strung on poles. For the New York system, he had ordered all the lines to be buried. The demonstration was a tremendous success, and the Edison Electric Light Company was off and running. Many orders came flooding in. By October 1886, the company had installed over five hundred generating plants across the country. Most, however, were isolated plants designed for a single building, not for a district.

## *A Curious Bind*

But Edison found himself caught in a curious bind. In the early part of the twentieth century, batteries were the only continuous source of electricity. Batteries produce direct current, and much of the development work on electricity, including Edison's, was based on this sort of continuous, steady current.

There was, however, a serious drawback to direct current (DC), one that only became obvious when Edison's Pearl Street Station showed that electricity could be a usable, practical source of power, not just for lighting but for other applications as well. In choosing to base his system on DC, he was backing the wrong horse.

Edison had made a strong selling point of the inherently greater safety of his electrical system over gas lamps. But this safety only held, he maintained, when the voltage was kept below about 500 volts, which was the case with his system. Even if someone were to accidentally come in contact with a line carrying this level of voltage, it would hurt but not kill. This fitted in well with the fact that DC actually worked best at these lower voltages.

The problem was that at these voltages, the effective reach of the power (from generator to user) fell off drastically beyond a distance of half a mile or so. In an urban setting like downtown New York, this was not a drawback. But for more isolated applications, it would make more sense to set up a large, central generator, perhaps located near waterpower or a source of fuel such as coal, and then "ship" the electricity to the outlying regions.

## *Alternating Current*

This could be done far more easily with alternating current, which turns out to be quite a different animal. While direct current is the steady flow of electrical energy in one direction, alternating current is regularly reversed in direction. Each complete reversal is called a cycle.

The easiest way to understand the distinction between these two forms of electrical current is to think back to 1831. In that fateful year, Michael Faraday, by that time very interested in electrical effects, tried the following. He set up a coiled wire, the two ends of which were connected to a sensitive meter that would detect and show a flow of current. He then thrust a simple bar magnet through the center of the coil. The meter pointer moved, indicating a flow of electricity in a given direction. When the movement of the magnet stopped, the meter moved back to the zero point.

When Faraday pulled the magnet out of the coil, the meter again showed electrical activity, but in the opposite direction.

He reasoned, correctly, that any relative movement between a magnet and an electric circuit would generate electricity. Moving a magnet in and out of a coil of wire would clearly be an awkward way to produce a continuous alternating current, but it soon became clear that a better way would be to rotate either the coil or the magnet, which would automatically produce a continuously varying current. This approach led eventually to the basic dynamo, commonly called an electric generator today. Though this method for generating current is basically the same for DC and AC, the way the current is drawn off is different.

Simple electric devices, such as lamps and electric heaters, operate equally well under DC and AC. With the initial concentration on electric lighting and urban centers, Edison's installations were perfectly satisfactory, and he built up a strong business installing them.

But that very success sowed the seeds of disaster. The promise of electrical lighting, and of using electricity for a variety of other applications as well, began to show the disadvantages of Edison's DC systems. Electrifying the entire business district in Manhattan would have required almost three dozen central generators, with the associated noise, haulage of coal and ashes, plus the smoke and smell of the burning coal.

Clearly, it would be better to create a large central generating station, away from the dense population, and then ship the electricity to the needed areas. With AC, and voltages in the multithousand range, the current could in fact be sent hundreds of miles over wires and then reduced to safer levels at the user's stations. At first, there was no device that could do this. But Faraday and Joseph Henry had also shown that voltage could be increased (stepped up) or decreased (stepped down) by transferring it between two coils, each with a different number of windings. This device would be called a transformer. If the primary winding has twice as many coils as the other, the secondary voltage will be half as large. But the effect only occurs when the current is turned on or off. In DC, this happens only once; with AC, it is a continuous process, and the step-up/step-down change is much easier to accomplish.

In the early years of Edison's business, no such device was available commercially, and Edison continued to lead the field. There was some competition from firms that generated electricity for street lighting, including arc lighting, but Edison remained far ahead.

But the idea that AC could be shipped wouldn't die. In 1883, Lucien Gaulard of France and John Dixon Gibbs of England patented their version of a transformer for an AC system.[5] It was an important step, but was not, by itself, enough to create big waves. Someone else, however, had become interested in the idea.

## *Tesla*

After Nikola Tesla emigrated to the United States from his native Croatia in 1884, at the age of twenty-eight, he was to play a brief but important role in the AC/DC drama. A brilliant if somewhat erratic electrician and inventor, he seemed ideal for employment by Edison, who was continually on the lookout for creative assistants.

At first, Tesla was assigned to travel around the various Edison manufacturing plants in Europe as a kind of roving trouble-shooter, and all went well. But when he began working as an assistant to Edison himself, here in the United States, differences began to show and cause trouble. That Edison was a trial-and-error researcher while Tesla was a man of theory was not in itself a real problem; Edison managed to work with others who were theoreticians. But these other theoreticians were more able to deal with Edison's tendency to belittle many theoretical ideas as impractical.

The relationship between Edison and Tesla lasted less than a year and fell apart over an apparent misunderstanding regarding a bonus that Tesla believed had been promised to him for certain work he did, but which Edison refused to pay.

There was an even more basic problem. Tesla had already planted some seeds in the garden of alternating current and had figured out designs for equipment that would work with this form of electricity. This included transformers that could accomplish the job of stepping up and steppping down voltage at the beginning and end of the transmission line.

Edison was just not interested in this approach. After Tesla left Edison, he was quickly approached by one of Edison's major competitors, one George Westinghouse, who hired him at a good salary. More important, Westinghouse purchased the rights to Tesla's designs for both motors and transformers.

## *Patent Fights*

By 1887 Edison had outgrown the famous laboratory he had created at Menlo Park eleven years earlier and had established a much larger headquarters in neighboring West Orange, New Jersey. In the meantime, however, starting around 1880, several other filament lamps, plus devices to support the lamps, had been developed and marketed by other groups. Edison had chosen not to institute legal proceedings because he wanted to devote his time to his inventions, and knew from other experience that such lawsuits would eat up vast amounts of his precious time. He was also very skeptical about the patent laws and truly felt he would not be able to obtain justice in that realm.

One of these lamp entries was a filament type developed by William Sawyer and Albon Man, for which they filed a patent claim early in 1880. Edison managed to keep it bottled up in the patent office for over five years by claiming it infringed on his own patent. But on May 12, 1885, the Sawyer-Man patent was officially granted. It was either bought or licensed by several of Edison's competitors.

Edison knew he had to open up a new front. Between 1885 and the end of the century, he instituted more than two hundred lawsuits—in the United States, Canada, England, and elsewhere—against what he considered infringers on the Edison electric lighting system. Roughly half of these suits involved the incandescent lamp itself.

But the situation was complicated by the fact that developments in other countries could affect his suits in the United States, and at first his skepticism regarding the judicial system and patent law seemed to be borne out. Suits he had brought in Canada and England went against him, and these affected his suits here. But in mid-1889, things began to change. The Canadian decision was reversed by the Canadian Minister of Agriculture, which allowed him to reinstitute a suit against United States Electric Lighting, one of his competitors.

In another patent battle, Consolidated Electric had brought a countersuit against the Edison interests based on their Sawyer-Man patent. But this suit, too, went in Edison's favor. In England, an appeals court had overturned an earlier decision against Edison. By 1900, then, Edison stood triumphant. But his legal costs had climbed to some $2 million, an astronomical sum at the time.

## *Westinghouse*

Watching all this very carefully was George Westinghouse, an American inventor and entrepreneur. Born in Central Bridge, New York, October 1846, he was just a bit older than Edison. Like Edison, his career started early, and by the age of twenty-two he had invented an air brake for railroad cars that was far more effective than the old mechanical ones. Building on that, he established large plants in Schenectady and Pittsburgh.

Edison's relationship with Westinghouse ran a roller-coaster course. Westinghouse had visited Edison at Menlo Park in 1880, and they apparently respected and even liked each other. But a year later, when Westinghouse decided to go after a contract involving a high-speed steam engine, Edison heard about this and suggested that Westinghouse knew nothing about engines and should stick to his air brakes. The comment got back to Westinghouse, who had, perhaps unbeknownst to Edison, already been working with not only engines but electric generators as well. A few years later, in 1885, Westinghouse decided to move strongly into this area and created the Westinghouse Electrical and Manufacturing Company, devoted to designing, building, and installing electrical systems.

He also bought control of United States Electric, which held the Sawyer-Man patent, and entered the electric lighting field in 1886. At this point, he had become a full-fledged threat to Edison.

It's likely that the personal competition, which was to loom large later, began here. In contrast with Edison, who was almost pathologically driven to invent and develop everything he needed himself, Westinghouse, although a prolific inventor himself, was quite content to buy up others' patents and put them to work in his organization. We have seen that he already had purchased some of Tesla's new work in the field of AC.

With these and his own inventions, he had built up an organization along the lines of the Edison interests, combining invention and manufacturing into highly successful enterprises. And just as Edison had done with a direct current electrical system, Westinghouse and his coworkers came up with a basic, workable electric system, but one that used alternating current. As with Edison's system, this involved all the ancillary equipment needed to make it work, including transformers, meters, motors, more-reliable generators, switches, and the like.

Westinghouse also included an AC generator developed by William Stanley, an American engineer and inventor. Stanley had first offered to sell the rights to the device to the dominant Edison. When Edison, still married to his DC system, turned him down, Westinghouse was happy to snatch it up.[6]

Using Stanley's transformer, Westinghouse set up a demonstration lab at Great Barrrington, Massachusetts, in March 1886, whereby he lighted several stores in the village. The demonstration included the generation of power at a central station, stepping up the voltage for transmission over a long length of wire, and finally stepping it back down to a safe level for lighting a group of prewired lamps in the village. The demonstration was successful, and a larger installation was set up in Buffalo. On the eve of

Thanksgiving 1886, that city became the first to receive power and light via a Westinghouse system.

Other demonstrations followed, and orders began to pour in. Within the first two years of operation, the Westinghouse system was installed in 130 towns and cities; his staff size leaped from hundreds to thousands. The Westinghouse system not only was able to transmit power efficiently over great distances, but was also cheaper to run and maintain than the Edison systems.

At first, the Edison people, believing that their lock on the production of light and power was impregnable, were unworried. But as the competition increased, pressure began to grow from Edison's own agents to move to, or at least work on, an AC system. Some of them even deserted to Westinghouse's side.

## *A Blind Spot*

But, in a rare instance of technical blindness and obstinacy, Edison held out long after he should have capitulated. So brilliant in other areas, he was simply wrong here. How could this have happened?

First, as a leader and originator in the power generation and distribution field, he had moved ahead strongly and quickly. Having built a large and successful enterprise, with a heavy investment in DC, he was loath to jettison almost everything he had done.

Second, his major competition was led by Westinghouse, and capitulation would be an admission that Westinghouse had, in a sense, bested him. This he found very hard to do.

Had he swallowed his pride and moved to AC earlier, he might well have been able to outcompete Westinghouse in the business arena. But as the Westinghouse organization grew, so did the conflict between them. And when Westinghouse decided to challenge Edison in the New York market, the fat was really in the fire. Edison, finally beginning to realize that, in truth, he could not win in either the economic or the technical sphere, sought another route, and their battle took a most unexpected turn.

## *Publicity Campaign*

Many of Westinghouse's installations, located in outlying areas, had the wires strung on poles aboveground, as was typically done with telegraph lines. Edison's side decided to challenge the AC method on the grounds that its use of high voltages was a menace to public safety. He began a major campaign in the public media, including articles in newspapers and general magazines, as well as lectures, pamphlets, lobbying, and anything else he could think up.

There was some precedent for a media-centered approach. Earlier, at the Paris Exposition of 1882, Edison had been able to successfully work the press for excellent coverage, at a time when his organization's electrical preeminence was being severely tested in the marketplace. He also had used the public safety issue very successfully when coming head-to-head with the entrenched gas lighting interests.

Considering his tendency to shy away from the scientific/theoretical realm, plus his earlier success in working with lay media, it isn't too surprising that he again turned to the public press rather than to technical and scientific journals in his attempt to bring public attention to bear on the putative dangers of alternating current. There had been some accidental deaths from electricity, and Edison's side tried to parlay this into a full-fledged fear of AC among the public. By a lucky bit of coincidence, the New York State Legislature, seeking a more humane method of execution than hanging, had begun to look into electrocution as a possibility. The legislature asked Edison for an opinion. At first he demurred, pointing out that he was against capital punishment.

Early in 1888, however, Harold P. Brown, a New York engineer, showed up at Edison's lab and said he agreed with Edison as to the dangers of AC. He offered to demonstrate these dangers in a series of experiments, and also to look into the suitability of AC as a mode of execution. Edison accepted the offer.

Working with Arthur E. Kennelly, one of Edison's most trusted electricians, Brown set up a series of experiments, in which no fewer than fifty dogs and cats were electrocuted—using AC generators purchased sub rosa from the Westinghouse organization. The press were invited and were happy to make hay from the grisly demonstrations. When critics objected that these animals were so much smaller than humans and didn't really show electricity's ability to execute humans, the Edison group upped the ante and demonstrated on a calf and then a horse!

Using these experiments as a springboard, the Edison group ratcheted up their publicity offensive against AC, pointing out how good it was at killing living things. Their articles also reported on other human deaths that had been occurring and argued that alternating current should be limited to no more than a few hundred volts.

Recall that Edison's initial response to the legislature's request for consultation on the use of electricity on condemned prisoners was that he wanted nothing to do with it. Yet he not only eventually went along with the demonstrations, but also allowed Brown and others, as well as himself, to make statements that were totally misleading.

Brown, for example, stated in a newspaper article, "It is well known that a continuous current of low tension [i.e., low-voltage DC], such as is used by the Edison Company for incandescent lights, is perfectly safe as far as life risk is concerned. . . . It is the rapid succession of shocks [in AC] that kills."

Later in the same article, he states, "There is only one excuse for the use of the fatal alternating current, and that is to save the expense of the heavier copper wires required by the safe incandescent systems. . . . In plain words, the public must submit to constant danger from sudden death in order that a corporation may pay a little larger dividend."[7]

To strengthen his charges, Brown issued a challenge to Westinghouse. Starting at 100 volts, Brown would take DC through his body, and Westinghouse would take a similar level of AC through his. Then the voltage would be upped 50 volts at a time, until one of them threw in the towel. Westinghouse, of course, ignored the challenge.

Not everyone was fooled. After an interchange between the two sides, perhaps at a debate, an editorial in the Boston *Post* on July 11, 1889, called Brown a warm, personal friend of Edison and stated, "There is something here that demands explanation. . . . The force of advertising could not go much further. . . . The testimony for the advocates and users of the continuous [direct] current was of, course, to the effect that their rival's method would kill a man quicker than a person could say 'Jack Robinson.'"

But, the *Post* article continued, "when the other side began their story, the whole face of the matter was . . . changed and confounded."[8]

The other side was, of course, Westinghouse and his AC systems. Westinghouse, no less a fighter than Edison, chose not to wilt under these withering attacks. One route was to play Edison's game, and to respond in kind—to spell out his position in the public press as well. In the *New York World* for March 18, 1888, he wrote:

> It is generally understood that Harold P. Brown is conducting these experiments in the interest and pay of the Edison Electric Light Company, [that] the Edison Company's business can be vitally injured if the alternating current apparatus continues to be as successfully introduced as it has heretofore been, and that the Edison representatives, from a business point of view, consider themselves justified in resorting to any expedient to prevent the extension of the system.

> To show the absurdity of connecting the [animal] experiments made at Mr. Edison's laboratory at Orange, New Jersey, with the commercial use of electrical currents it is only necessary to call attention to the manner in which the current was applied. . . . The method of applying the current . . . was carefully selected for the purpose of producing the most startling effects. . . . The parts . . . were so carefully placed in the circuit as to receive a shock such as would be utterly impossible if the current were applied in any ordinary or accidental manner.

[There follows a discussion of the varying resistance offered by different bodies, and how this, too, enters into the equation. Then he continues:]

> It was for this reason that a vital part of the body was selected. . . . [A] large quantity of current was made to pass through sensitive portions of the brain and spinal cord . . . [whereas] the usual points of contact in accidental shocks received from electrical circuits are through the hands, or some other portion of the body protected by tissues. . . .
>
> We have no hesitation in charging that the object of these experiments [on the animals] is not in the interest of science or safety, but to endeavor to create in the minds of the public a prejudice against the use of the alternating currents.[9]

Westinghouse then goes on to give figures on the strong sales figures of AC versus DC systems; and also explains that the higher voltages are used only for transporting the electricity to distant or out-of-the-way regions, and that these high voltages are then "stepped-down" to 50 volts or so before being fed into homes and offices, where people were more likely to come in contact with the wires and other electrical equipment.

Using any route they could come up with, the Edison group also began lobbying for a law to limit all electric circuits to a few hundred volts, which would, of course, effectively kill AC.[10]

Westinghouse's side accused his opponents of conspiracy. And so on.

Despite strong protests by Westinghouse that his equipment was being used unfairly, the New York prison authority decided to try out this new method, and after many delays, proceeded to execute William Kemmler, a condemned murderer, on August 6, 1890, using AC. Unhappily, the level of charge sent through his body was inusufficient, and the awful process had to be done over.

The electrocution had hardly been painless and more humane than hanging. Ignoring this inconvenient fact, Brown wrote, "In respect to the

Kemmler case, I have no doubt that the Court of Appeals will sustain the Electrical Execution Law, for I cannot see how Referee Becker can do otherwise than report the Westinghouse machine as an instrument of instantanous and painless death."[11]

Perhaps the most ingenious move was an attempt to connect Westinghouse directly and forever with electrical execution. An unsigned article in the Buffalo *Courier* for August 3, 1889, contained this statement:

> A gentleman prominent in electrical circles [almost surely Brown] said that there is one word which, in his judgement, might be used under the circumstances with some propriety. . . .
>
> As Westinghouse's dynamo is to be used for the purposes of producing the alternating current to execute criminals, why not give him the benefit of this fact in the minds of the public, and speak hereafter of a criminal being 'westinghoused,' . . . or being condemned to the westinghouse. It would be a subtle compliment to the business enterprise of a well-known man.[12]

Fortunately for Westinghouse, the description never took hold. But as we know, electrocution—handled more efficiently—was indeed to become the preferred method of capital punishment. It remained so for almost a century and only began being replaced by lethal injection in the 1970s.

Recognizing the futility of this public jousting, Westinghouse put out feelers for some sort of settlement of their feuds, perhaps even a merger. Edison would have no part of it. He argued that Westinghouse's "methods of doing business lately are such that it cannot be accounted for on any other grounds than that the man has gone crazy over sudden acquisition of wealth, or something unknown to me, and is flying a kite that will land him in the mud sooner or later."[13]

Edison may also have begun to reconsider his position on AC. By October 1890, Arthur E. Kennelly was taking a closer look at AC systems. Perhaps Edison felt that if push came to shove, he could eventually come up with something better and overpower Westinghouse in the marketplace.

## *Counterattack*

In any case, Westinghouse saw that he needed to change his strategy and had to mount some sort of strong counterattack. But it would have to be one that commanded attention—some sort of powerful demonstration. He had long thought about harnessing the incredible power contained in the water flowing over Niagara Falls, but a more immediate opportunity presented itself.

The sponsors of the upcoming Chicago World's Fair (also called the Columbian Exposition, scheduled for 1893) wanted to show the world just how far the United States had come since Columbus had set foot on the American continent four centuries earlier. And what could do this better than lighting up the fair as no city had ever been lit up before? Edison and Westinghouse were the only bidders, and their battle burst forth in renewed fury.

On one side stood the Edison General Electric Company, a powerful organization that had absorbed and combined several of Edison's manufacturing companies in the spring of 1889. Formation of this company showed, however, that a small revolution was in process. Edison had long been used to being in charge of just about everything and keeping his finger in every pie; but the size and complexity of his companies was such that his lack of formal education was catching up to him. He also needed to raise funds to keep up with the needs of his invention factory. Edison received 17.5 percent of the stock in the new company and separate financial help in the costs of his laboratory expenses, which were considerable. He remained active in the new company but was beholden to a consortium of financiers, including the powerful J. Pierpont Morgan.

Edison General Electric held the patent rights to Edison's heated-filament incandescent lamp, and management felt that they therefore held a trump card. Their bid was $14 per lamp for the 250,000 lamps that were to light up the fair.

But Westinghouse Electric had a powerful trick up its sleeve, and the Edison group was dumbfounded when Westinghouse offered a bid of $5.25 per lamp.

The Edison group, having obtained an injunction against the Sawyer-Man lamp as too similar to Edison's, may even have thought that Westinghouse could not bid at all. But Westinghouse had thought ahead and had had a lamp designed that got around the Edison patent. Called a stopper lamp, it wasn't as good as Edison's, but it did the job for the short run of the fair. Westinghouse knew his group would make no money on the low bid, and would probably even lose on it, but he felt it was worth doing.

With the promised savings of over a million dollars, the fair sponsors had little difficulty in awarding the contract, on May 23, 1892, to Westinghouse Electric.

The Columbian World Exhibition opened on the morning of May 1, 1893. Among its many wonders was the Westinghouse electrical system. The largest such system built to that time, it starred twelve 75-ton dynamos, with much of the generating and distribution system based on Tesla's earlier work with alternating current. But it was at night that the fair really earned its title of Magic City.

Westinghouse's lighting demonstration was a resounding success. By this time, however, Edison's interests were moving elsewhere, and the world had changed even further for him.

By the year 1890, there were basically three major players in the lighting and electric world. These were the Edison General Electric Company; the Consolidated Electric Company, which held the Sawyer-Man patent and was controlled by Westinghouse; and Thomson-Houston, which like Westinghouse Electric had been infringing on the Edison patent and had built up a strong business in the arc lighting world. In addition, however, it was also moving strongly into the AC field. It was already far ahead of Edison in that respect.

At the same time, thanks to a serious bank failure in London, money for investment was becoming harder to come by. The directors of the three companies began looking for ways to ease up on the patent squabbles, and perhaps to merge somehow. The wealthy stockholders in Thomson-Houston were on good terms with the Morgan group now running Edison General Electric. But when Edison was approached with the idea of combining with Thomson-Houston, he immediately turned it down. He argued that "having boldly appropriated and infringed upon every patent we use, there is very little left to favor them with, except our business, which they are now after. . . . The more I figure the 'benefits' of a coalition the more worried I am that you may be induced to enter into one. . . . If you make the coalition my usefulness as an inventor is gone."[14]

Nevertheless, and after much wrangling, Thomson-Houston and Edison General Electric were merged in an agreement dated April 15, 1892. But this time the company was to be called General Electric—Edison was taken off the marquee.

So it was that when bids were submitted in March 1893 for the job of electrifying Niagara Falls, they came from Westinghouse Electric and from General Electric. Again, Westinghouse prevailed, and the primacy of alternating current was clearly evident.

## *The Later Years*

Although Edison had actually been removing himself slowly from the running of the newly formed company, there is no doubt that he felt the expulsion of his name very strongly. His secretary, Alfred Tate, later reported, "Something had died in Edison's heart."[15]

Edison maintained that he was done with electric lighting, that there were other areas he wanted to delve into, and that he now had the time and money to devote his attention to them. During the 1890s, he did some work on motion pictures and on improving his phonograph. Mainly, however, he spent a huge amount of time, energy, and money on trying to develop a kind of magnetic ore separation process, which did not work out.

During that period, he spent much of his time in the field, and the West Orange laboratory was reduced to a skeleton crew. In the late 1890s, after giving up on the ore work, Edison returned to his beloved laboratory activities.

He began work on a better way of manufacturing Portland cement, using some of the experience and equipment he had developed in the ore milling process; and he began work on a nickel-iron, alkaline storage battery, which offered some advantages over the more common lead-acid battery.

In 1911, he reorganized his remaining companies into a single group called Thomas A. Edison, Inc., which was to be devoted to his new inventions. He still had an active business based on these inventions. Though Edison's battery never found a home in electric cars (the advent of the electric self-starter gave gasoline-powered cars a head start they never relinquished), it found wide use in many other ways, including a very successful miner's lamp that provided a significant increase in safety; lights in railway cars for railway signaling; in trucks; and for remote lighting purposes in rural homes and lighthouses. His battery, in fact, became one of the most successful products in his long inventing career.

Edison remained active almost till the last of his days. Among other activities, he played a large part in the early development of motion pictures; he spent some time and energy in trying to get a research laboratory going for the U.S. Navy, which did not happen; and he tried to find a domestic source for natural latex rubber when a shortage seemed imminent.

He died on October 18, 1931, having left a legacy that few, if any, can match. Almost single-handedly he had transformed the domain of invention from that of the individual tinkerer to that of the organized industrial research group. He also showed that support for industrial research could be a wise investment over the long run. He was, however, a generalist who lived long enough to see, though probably not recognize, another change taking place—namely, that the major players in the world of industrial research, and of increasing specialization, were coming more and more from the ranks of the well-educated scientists and engineers, who could bring strong analytical and mathematical skills to bear.

Westinghouse, too, found himself being bypassed by his own organizations. In 1907, a widespread money crisis left several of his main businesses in serious financial trouble. One outcome was a reorganization in 1908, in which he began to lose control. By 1911, he, like Edison a decade earlier, was no longer in charge. Though he was no longer a major player in running the companies he had helped form, his name lived on and continued to be one of the organization's major assets.

Like Edison, Westinghouse had been a prolific inventor, with well over three hundred patents granted by the time he died on March 12, 1914. More important in his case was that over the course of his long career, he had established close to seventy separate businesses. In contrast with Edison's companies, which often involved outside investors (but which were not publicly traded), Westinghouse generally did his own financing, often securing capital for a new enterprise by personal appeals to friends and stockholders. Because he put so much of his money back into his enterprises, he did not leave the expected fortune at the time of his death.

Ironically, even though both Westinghouse and Edison were no longer involved in the AC/DC-current war, the electrical companies that carried on after them still found themselves in competition now and then. The huge turbogenerators that began driving out the smaller generators in many of the smaller, isolated plants were being offered in different configurations by both organizations. Westinghouse favored horizontal shaft units; General Electric favored a vertical shaft.

On the other hand, the two organizations managed to complete several cross-licensing agreements involving electrical equipment that seem to have worked out satisfactorily.

Recall George Stephenson's prophesy—that electricity would one day be the main power source all over the world. For that remarkable accomplishment, we can thank both Edison and Westinghouse.

Remarkably, although AC eventually was adopted as the best route to widespread electrification, it has still not completely replaced DC. Direct current is in use today for equipment that may have been installed as late as the 1920s and continues to operate. Examples include elevators, water pumps, and some rapid transit lines. As these items wear out, however, they are almost invariably replaced by AC equipment. Direct current also also finds use in a variety of specialized laboratory and chemical applications. In all these cases, the power is brought to the applications as AC, and is then converted, or "rectified," to DC.

And even though both men lost control of their electrical enterprises in later years, this was basically because of their huge success; that is, because of the huge organizations these businesses had become. Nevertheless, the names Edison and Westinghouse remain household words, and their memories are honored in any history of technology.

CHAPTER 5

# Ford versus Selden and ALAM

## Automobile Manufacturing

In the spring of 1903, two organizations were born that were to revolutionize the meaning of the word *manufacture*—but not before a pitched battle that raged for eight long years. At the end of it, the Ford Motor Company and Henry Ford—the small-town boy from the Midwest—stood triumphant over entrenched interests—the Association of Licensed Automobile Manufacturers (ALAM) and George Selden—in the hated East.

Had Ford lost, it's very likely that these interests would have squelched a budding industry—the American automobile industry—which in a few short years sparked the engine of American industrial might and changed the face of manufacturing forever.

In the process, Ford became one of the most admired people in the world. Robert Lacey, one of his biographers, describes him as "innovator, giant-killer, and tribune of the people."[1]

It's a story that involves monopoly and strong-arm tactics on Selden's side; and publicity, duplicity, and bravery on Ford's side. It starts a quarter of a century before 1903, when a clever patent attorney saw a way to make a killing.

### The Beginnings

The *idea* of a motorized conveyance had been around for a while. In 1863, the same year that Ford was born, Sylvester Hayward Roper of Roxbury, Massachusetts, had actually built one. It was, in truth, not much more than a buggy powered by a small, but heavy, steam engine. Two years later, he had improved the design of the engine. His buggy now included a two-cylinder charcoal burner, and was probably the first operable motor vehicle in the United States—though it had no brakes. The driver stopped by throttling down, or engaging reverse.

By the 1890s, steam cars were common; and one, the Doble, continued to be sold until the 1930s, even though the internal combustion engine had by that time surged well past steam and electricity as the favored power source for autos.

But in Roper's time, the horse was king, and his Rover was regarded more as a curiosity than as a potentially important mode of transportation. For one thing, the public wondered about the wisdom of bumping along while sitting above a boiler filled with live steam. Such boilers had the reputation of exploding now and then. They were also expensive and hard to operate and control.

A dozen or so years later, a German experimenter, Nikolaus Otto, came up with a workable gas-powered internal combustion engine. In modern terms, internal combustion merely means that the fuel is fired inside the engine, rather than externally as in the steam engine. The power actually comes from small, gasoline-powered explosions that drive a piston back and forth inside a cylinder. Otto's engine, called a compression engine, wasn't quite there, but it came close, and provided a major stepping-stone for Henry Ford. Earlier attempts at putting gaseous or liquid fuels to work by a variety of inventors were more of a cross between the steam engine and a modern gasoline engine. The application of the Otto engine to the automobile would come eventually, but not quite yet.

## *Henry Ford*

Henry Ford was born on a farm near Detroit. By the mid-1870s, when he was in his early teens, it was clear that he was no scholar; in fact, he never lost his dislike for books. He also disliked horses and farmwork in general—not useful traits for a farmer. What he did love was machinery, any kind, from watches to engines.

"Often," he wrote later in his autobiography, "I took a broken watch and tried to put it together. When I was thirteen I managed for the first time to put a watch together so that it would keep time. . . . There is an immense amount to be learned simply by tinkering with things. It is not possible to learn from books how everything is made—and a real mechanic ought to know how nearly everything is made. Machines are to a mechanic what books are to a writer. He gets ideas from them, and if he has any brains he will apply those ideas."[2]

"From the beginning," he goes on, "I never could work up much interest in the labour of farming. I wanted to have something to do with machinery." He also refers to the "excessively hard labor of ploughing," and he thought often of replacing horses with machines. In other words, his first interest in putting motors to use was not for transportation but rather to lighten the farmer's load.

He continues, "The biggest event of those early years was meeting with a road engine."

Road engines at the time were usually steam-powered contraptions that would be set up at a farm to drive farming equipment, such as a thresher, or to power, say, a sawmill. But this road engine, Ford points out, "was the first vehicle other than horse-drawn that I had ever seen. . . . I had seen plenty of those engines hauled around by horses, but this one had a chain that made a connection between the engine and the rear wheels of the wagon-like frame on which the boiler was mounted.

"From the time I saw that road engine as a boy of twelve . . . my great interest has been in making a machine that would travel the roads." He later built a workable steam-driven car. But after two years of part-time experimenting, he dropped the idea when he saw that because a steam-based engine required very high pressure, it must be very strong, and hence heavy, to make it safe.

After putting in some time as a machine shop apprentice, he worked as a local representative of the Westinghouse Company, setting up and repairing their road engines. By this time, around 1883, these devices were often self-propelled, and could make about 12 mph on the road. But they were very heavy, typically a couple of tons, and very expensive. This bothered Ford, and he left after a year.

After other odd jobs, his father offered him some forested property, and in 1886, Ford moved back to the land and did some logging and farming. While there, he took on some jobs in nearby areas as a sort of peripatetic mechanic and repairer of broken machinery. He also set up a workshop on the farm and did some of his early development work on the car in that shop. In the meantime, he had been following the progress of the Otto engine, with which he already had some experience. In 1885, he had repaired one, and in 1887, he built one.

But internal combustion engines require electricity to power the tiny explosions, and Ford felt he needed more training in this area. In 1891, he obtained a position as mechanic/engineer, overseeing the electric generators at a substation of the Edison Illuminating Company in Detroit.[3]

In 1896, he met Thomas Edison, and they later became fast friends. Ford, in his autobiography, describes Edison as "the world's greatest scientist."[4]

His admiring evaluation reflects an interesting difference between the two men. Edison, you recall, sought to develop or invent every device he ever manufactured and/or sold. Ford had no such ambitions. His genius lay in applying new ideas no matter where they came from. By 1896, in an experimental shop he set up in a woodshed in back of his house, he built a tiller-steered contraption mounted on bicycle wheels.[5] His Quadricycle was powered by an internal combustion engine of the Otto type—one that Ford had built himself. It was his first entry in the automobile sweepstakes, and the date of its completion was to play an important part in the battles to come.

## *George Selden*

George B. Selden, like Ford, had a strong mechanical capability; but just about everything else about their backgrounds was different. Born in 1846, he came from a well-to-do family in the city of Rochester, New York, and attended Yale University. His father, a judge, wanted him to go into law. He compromised by becoming a patent attorney and establishing a practice in Rochester. He practiced law to please his father, but he also dealt with his own love, mechanical devices. He had long been thinking about motorized conveyance and had done some experimenting in his spare time. From 1873 to 1876, he spent many hours in his basement laboratory, testing the characteristics of different engine fuels. He, like Ford, felt that steam-powered vehicles had too many disadvantages, and he began working on a possible alternative.

But this was before the Otto engine had become well-known. In 1876, Selden had attended the Philadelphia Centennial Exposition, where there were many engines on view, including a design by George Brayton, an American engineer. An early forerunner of the diesel engine, it compressed the fuel charge outside the cylinder, and burned the charge rather than exploded it. An early entry into the gasoline-engine world, it did burn atomized gasoline, but was still massive, and sat in a bed frame some ten feet long. Selden felt, however, that it could be pared down in size and weight, and that it might serve as the basis for his dreamed-of conveyance.

In early 1877, he was busy with his patent practice; although he could only work on his ideas intermittently, he did manage to sketch an engine driving the shaft of a road vehicle. By the end of the year, he had set some specifications for a slimmed-down Brayton engine, and he enlisted an experienced machinist, Frank H. Clement, to build a multicylinder gas-powered engine based on the Brayton design. By 1878, they had pared down the 1.4 horsepower Brayton engine from 1,160 pounds to about 370 pounds, but had only managed to fabricate one cylinder. With all three cylinders working, Selden's design would have produced 2 horsepower. At the time, it should be noted, the Otto engine was also a massive device.

So Selden had made some progress, but he saw that with part-time tinkering, it would take forever to complete work on the motor and incorporate it into a vehicle. He decided to put his patent experience to work.

In 1878, he had a draftsman do some drawings of a vehicle and engine, and got another machinist, William Gomm, to create the model required by the Patent Office. It was not a working model, but Selden hoped he could get away with presenting "in outline the general features" of a motor car powered by a Brayton-type engine. He filed his application in the spring of 1879.

## *A Clever Plan*

His patent has been described as covering a "self-propelled vehicle comprising steering wheel, a liquid hydrocarbon engine of the compression type with the engine running at a speed greater than the driven wheels, a disconnecting means between the two, and a body adapted to either persons or goods. In effect, these were rather broad claims, written in terms of technical categories rather than specific devices."[6]

The patent was granted, but not until sixteen years later, in 1895. The long hiatus was the best thing that could have happened to Selden—and was, in fact, part of his plan. For by 1895, the automotive world had changed enormously. In the mid-1880s, two German engineer/inventors, Gottlieb Daimler and Karl Benz, had come up with designs for a working gasoline-powered motorcar. Then Daimler built and patented the first high-speed internal combustion engine, which was lighter and more efficient than Otto's (and others') low-speed engines. He also came up with the carburetor, which meant the engine could run on liquid gasoline rather than a gaseous hydrocarbon, which was much more expensive to operate.

A year later, Benz obtained a patent on a vehicle that was powered by a four-cycle engine he had designed and built the previous year, the same time that Daimler's engine came into being. The four-cycle engine remains the standard today. Its advantage is that all the processes—fuel intake, compression of the gases, ignition, and exhaust—take place within the cylinder in rapid succession.

Some say that Benz's vehicle was the world's first practical conveyance powered by an internal combustion engine. Others maintain that credit for that development probably should be divided between the two men. In either case, much of the initial engineering work on the gasoline-driven auto was done in Germany. Though not described here, there were several major developments in France, as well.

There were a few foreign patents applied for in the United States, but in general the American market was wide open, and some of the patents had already expired. By 1893, the first American entry into the gasoline-powered automotive sweepstakes was constructed by the Duryea brothers, Charles and Frank. Over the next few years, many small manufacturers began trying their hand, and automotive development began shifting to the United States.

But see now how clever Selden's plan was. For as the new developments were taking place, he was able to include many of them in his patent by ingenious revisions in his claims. William Greenleaf, who did an extensive study of the Ford/Selden battle, describes Selden as "a consumate master of systematic and intentional delay."[7]

For example, Selden would draw up an amendment with one or more claims that he fully expected the Patent Office to turn down. This led to prolonged negotiations, thus extending the life of the patent. By the time the patent was issued, its history shows about a hundred separate changes, with all nineteen of the original claims replaced by updated revisions!

We see a good example of this in the case of the engine. Although Selden did not specify that he was basing his design on the then existing Brayton engine, it was clearly of that design. The patent as finally issued, however, covered the combination of the generalized features noted above with a compression-type internal combustion engine using liquid hydrocarbon fuel. If Selden had left his patent as initially worded, he would have had dominion only over motor cars using a Brayton-type engine. By his clever juggling, he was able to get it reworded to include all types of compression engines (another term for Otto's more advanced type of gasoline engine).

Selden was not the first to use this delaying technique—both Morse and Edison used it—but Selden seems to have been the master. He was helped by a combination of two difficulties faced by the Patent Office at that time. First, the patent rules permitted a long-drawn-out procedure. In addition, the Patent Office was being buried under an avalanche of records, sparked by the burgeoning Industrial Revolution, which left it understaffed and overworked.

Had Selden gotten the patent in 1879, it would have virtually expired by this time and he probably would not have profited at all. He might even have liked some further delay, but by the mid-1890s, the Patent Office began tightening up the rules that permitted these delaying tactics. This tightening may have had something to do with Selden's patent being granted at this time—sixteen years and six months after its first application.

As of 1895, none of the elements in his patent was new in design, nor did Selden claim so. He asserted that it was the combination of these features with a compression-type internal combustion engine using liquid fuel that was new.

In any case, Selden's patent now loomed large. In fact, it gave him a stranglehold on much of the budding American automobile industry. He had only to sit back and watch the dollars roll in. Every American manufacturer who wanted to build and sell a gasoline-powered motor car—and by this time there were many—would have to get a license from Selden.

But rather than do battle by himself, he chose to enlist the resources of a larger organization. In 1899, four years after Selden's patent was finally granted, he sold the rights to the Columbia & Electric Vehicle Company, a powerful Wall Street conglomerate with interests in several fields, including electrically driven streetcars, batteries, and cars, especially electric vehicles.

It's likely that the initial idea behind the purchase of Selden's patent by the conglomerate was to hold down competition with their electric vehicles. They realized very quickly, however, that they were backing the wrong horse, and the Electric Vehicle Company, a major part of the conglomerate, was doing badly. In their need for funds, they at first tried to turn the financial tide by taking on any company that wanted to license the Selden patent.

But rapid development in the world of internal combustion engines gave the group another idea. They would permit further development and sales of gasoline-powered vehicles, but would pick and choose among the manufacturers—and make the chosen companies pay handsomely for the privilege of being among the privileged group. They would use these funds to help shore up the Electric Vehicle Company.

## *A Growing Industry*

The only way to put teeth into their monopoly was to file infringement suits against trespassers. But lawsuits are slow and expensive, and in general, the industry continued to operate, and grow, unhindered. But it was largely a growth in numbers of manufacturers, who typically made only a few cars a year.

Yet there was a new idea in the automobile world: quantity manufacture. The first man to take that idea and run with it was Ransom E. Olds. Initially a builder of steam-driven vehicles, Olds decided that the future lay in the internal combustion engine. In 1897, he and S. L. Smith, a copper millionare from Detroit, formed the Olds Motor Company, which would develop and build gasoline-powered autos. But a problem arose almost immediately. Olds wanted to build and sell lots of cars at a reasonable price. His backers were more interested in putting out fewer but more expensive vehicles.

Olds's idea prevailed, and the company saw a strong initial success; by 1903, it had sold several thousand cars at $650, far outselling all other makes. But many of the early cars were unreliable, and the image of the frustrated driver lying under his car, trying to fix whatever had gone wrong that time, was all too common. A groundswell of irritation led to increased sales of better, if more expensive, cars, and the Olds company later lost its footing. But the company, and Olds himself, are generally given credit for making the first major move toward mass production of autos.

In the preceding years, Ford had gotten involved in two unsuccessful launches of motor vehicle manufacture, but had built some racing cars and had gained some prominence in that area. In fact, one of Ford's failures in setting up a motor company came about when he began work on another racing car, rather than concentrating on the manufacturing end. The investors pushed him out. He could take the designs for the racing car with him, but not those for the car he should have been developing. But, as with Olds, there was a difference in philosophy. He was thinking simple, inexpensive, and mass-produced; the investors had in mind luxury, as high a price as possible, and higher profit per car.

After the split, these investors investors started afresh, and in 1909, they created the Cadillac Motor Company. Named after Antoine Laumet de la Mothe Cadillac, a French army officer and explorer who played a major part in the founding of Detroit, their deluxe auto formed a stepping-stone in what was to become the automotive giant, General Motors. But the car was initially intended to be a Ford![8]

In those days, racing was actually one of the main ways of selling cars. The objective was to show how advanced the company was by building a car specifically for racing, or by backing one of its own models in races of various kinds.

Although Ford was not an inventor in the same class as Edison or Westinghouse, he was a master mechanic. Putting together a racing car in those days generally meant building everything from scratch. In 1901, Ford challenged Alexander Winton, a carmaker and racer who was among the leaders in the racing field.

For his entry, Ford designed a new electrical apparatus for the spark plug and got a dentist to make a porcelain case for it. This was the forerunner of the modern spark plug, still an important component of the ignition system. After an exciting race, Ford won.

## *The Ford Motor Company*

Ford's growing reputation in the racing area was an important factor in getting backing for his next, and most important, venture. By 1903, he felt he was ready to plunge deeply into the manufacture of gasoline-powered autos, and with the help of another group of investors, he launched the Ford Motor Company.

In general, the early auto manufacturers took orders and manufactured to order. Down payments, or even cash in advance, gave them cash to work with. Ford's idea was to manufacture in high quantity and sell from inventory, which led to a difficult cash flow at first. But the idea caught on, and the relatively low price of his first entry, the sturdy and dependable Model A, led to quick returns, and the company was well launched. By 1905, the Ford Motor Company had 450 agencies selling Ford cars across the country.

In 1906, the company introduced the Model N. Although Ford hoped to sell it for $450, he found they had to charge $600. This was not the cheapest car around, but it was a good car for the money. Ford's slogan was: "No car under $2,000 offers more."[9]

Lowering the price like that could be taken as strictly a business decision, in that it gathers more customers and thereby increases sales. But he deserves credit for a deeper, and more public-spirited, aim.

First, he honestly believed that he could make customers for his cars out of his workers. That was actually a novel idea. "The true industrial idea is not to make money," he maintained. "[It] is to express a serviceable idea, to duplicate a useful idea, by as many thousands as there are people who need it."[10] He did this by lowering the price of his product while at the same time increasing the workers' pay, in some cases doubling the going rate in the industry. This took some bravery and independence, for it certainly did not endear him to his colleagues in other companies.

He also had some interesting ideas on the industrialization of society.

> To produce, produce; to get a system that will reduce production to a fine art; to put production on such a basis as will provide means for expansion and the building of still more shops, the production of still more thousands of useful things—that is the real industrial idea. The negation of the industrial idea is the effort to make a profit out of speculation instead of out of work. There are short-sighted men who cannot see that business is bigger than any one man's interests. Business is a process of give and take, live and let live. It is cooperation among many forces and interests. Whenever you find a man who believes that business is a river whose beneficial flow ought to stop as soon as it reaches him you will find a man who thinks he can keep

> business alive by stopping its circulation. He would produce wealth by this stopping of the production of wealth.[11]

He certainly had Selden in mind when he wrote that.

On machines: "For most purposes a man with a machine is better than a man without a machine."[12]

On industrialization: "The extreme Socialists went wide of the mark in their reasoning that industry would eventually crush the worker. Modern industry is gradually lifting the worker and the world."[13]

Whether one agrees with these ideas or not, it is certain that Ford, virtually single-handedly, opened the automobile manufacturing field to all comers. He later put his feelings about competition and monopoly this way: "I found that competition was supposed to be a menace and that a good manager circumvented his competitors by getting a monopoly through artificial means. The idea was that there were only a certain number of people who could buy and that it was necessary to get their trade ahead of someone else."[14] He honestly felt that he could do some good as well as make some money.

In other words, he really did seem to have a social objective; and with the introduction of his next entry into the automobile world, he got his chance to put his ideas to work.

## *The Model T*

Ford set aside a room in his factory for developing his new ideas. The car that emerged in 1908, the Model T, was a wonder in many ways. Its fundamental characteristics were simplicity, reliability and durability—qualities desperately needed in those days of rutted, potholed, muddy and generally miserable roads. The car—suspended on indestructible leaf springs and rolling on light but strong chrome-vanadium steel axles—was an amazing success, and Ford continued to build them all way up to 1927!

The legend that the car never changed in all that time is not true. While the chassis and engine remained basically unchanged, there were many changes in equipment, materials and styles. The speedometer provides an interesting example. In 1915 Ford discontinued their use altogether because he found he could not obtain any that he considered reliable enough.

The only *significant* changes Ford made in the car over its amazing 18-year history were the addition of electric lights and an optional electric self-starter. By the early 1920s, owners who loved the car but wanted a few refinements could purchase them from a wide selection at a Sears, Roebuck store. That company offered an even wider selection of both parts and accessories in its Catalog of Automobile Supplies.

Though the Model T was a good car, with a number of improvements over anything running at the time, Ford's genius lay not in the car, but in the manufacturing processes that permitted him to keep lowering the price in the years that followed. Introduced in 1908 at $850, the car's price dropped to as low as $300 in the years that followed.

## *Challenge*

But in 1903, all of this was still a dream, for the Ford Motor Company was just getting started, and was dealing with cash flow problems. At the same time, Ford's opposition was throwing its weight around. Three years earlier, Selden and the Electric Vehicle Company (to which the Selden patent had been assigned) had brought an action against two important manufacturers, the Buffalo Gasolene Motor Company, a manufacturer of automobile parts, and the Winton Motor Carriage Company, which had become one of the leading auto manufacturers.

In short order, a group of other independents who felt similarly threatened came together with these two and circled their wagons for mutual protection. Arguing that the Selden patent was offering nothing really new, their lawyer filed a writ of demurrer to the district court requesting dismissal of the Winton suit. In November of that year, the court turned them down.

The next blow against the independents was struck when a district court judge in Buffalo, John R. Hazel, handed down a similar opinion. It overruled the demurrer requested by the independents for the Buffalo company.

Winton continued to battle, and the case went to court, with some financial help from the other independents. The case dragged on for two years and produced two thousand pages of testimony.

To no avail. The decision went against them. In the meantime, the Selden interests had instituted suit against other independents. At that point, Winton and the others gave in, and their association quickly withered away. They accepted the priority of the Selden patent and acquired the required license.

Not long after, and just a few weeks before the formal opening of the Ford Motor Company, the Electric Vehicle Company, which was the acting body in the suits being brought against the independents, joined with a group of the new licensees and formed the Association of Licensed Automobile Manufacturers, or ALAM.

Now, trade associations have a long history and can be useful to their members. But an association like ALAM, which was based upon a strong patent, then had the power to determine who was in and who was out of the gasoline-powered automobile business. And whoever was out, was out of business. The members of ALAM, most of which were firms that had spent time and money on invention and development, could feel the hot breath of competition from an ever increasing number of Johnny-come-latelies who chose not to acquire licenses.

At first, Ford figured it made sense to become part of ALAM. He called on Fred L. Smith, the treasurer of the Olds Motor Company and acting president of ALAM, and asked a hypothetical question: If Ford Motor Co. were to apply for membership, would it be accepted? Smith stated in a condescending manner that ALAM would probably not accept such an application and that he saw Ford's company as little more than an assembly plant. This was clearly a smoke screen, for none of the existing members manufactured all the parts themselves.

Another report puts the refusal on even more personal terms. It states that Ford "was firmly told that he was a poor risk; that as a person he was unfit for the responsibilities of manufacture; [and] that his flivvers were a disgrace to the dirt roads upon which they ran."[15]

Nevertheless, the Ford people, short of capital and wary of a prolonged patent battle, swallowed their pride and tried again. This time they were told that they were seen as a fly-by-night firm, and they should come back when they became a going concern.

ALAM, on the other hand, was growing rapidly. By the autumn of 1903, it had twenty-seven companies signed up.

No one knew it at the time, but ALAM's refusal was probably the best thing that could have happened to Ford's budding company. While the finally agreed-upon license fee, 1.25 percent of the retail price of any cars sold, did not sound like much, it could affect a purchaser on a tight budget. More important, with Ford thinking in terms of mass production and thousands of cars, the levy would add up to a considerable sum.

## *Ford's Fighting Instincts*

But most important of all, the rejection brought out Ford's fighting instincts and intensified his instinctual distrust of monopolistic behavior. In the forthcoming battle, he would challenge the very fiber of the patent system.

Thinking it over, Ford became convinced that the Selden patent would not stand if properly attacked. He undoubtedly also was buoyed by the public distaste for powerful trusts and conglomerates that was sweeping across the country. By 1905, and with Ford's urging, twenty of the outsiders, including Ford, came together and formed the American Motor Car Manufacturers Association in hopes, once again, that concerted action would help them counter the ALAM monopoly. The two groups engaged in a long series of judicial attacks and counterattacks.

Then ALAM effectively highjacked the all-important national automobile show, held annually at New York's Madison Square Garden. In response, the independents organized their own show, and by 1906, they could advertise that its group, the AMCMA, offered twice as many exhibitors as the competing show.

Then, something very curious took place. On June 14, 1907, Selden produced a working motor-driven vehicle, which he claimed he had been working on when he filed his patent. He felt this would provide additional evidence of his claims. Of course, it included some of the advances that were finally included in his patent as issued.

The vehicle was to be demonstrated in front of Judge Charles Merrill Hough, who was the sitting judge on the still-in-progress Selden case, at a racetrack near Guttenberg, New Jersey. After Selden started it with the aid of an air compressor, it ran some fifteen feet and died.

The judge—who was a hard worker but unversed in both patent law and automobile history—chose not to hold this outcome against the patent, and on September 19, 1909, he upheld the Selden patent. Probably influenced by Selden's use of highly paid experts to support his side,[16] Judge Hough was convinced that Selden had put together a number of preexisting elements and had come up with a new "harmonious whole capable of results never before achieved."[17]

ALAM, now in the driver's seat, intensified its actions and—in an extensive advertising campaign in both the Detroit newspapers and in some automotive trade publications—even threatened *purchasers* of gasoline-powered vehicles made by non-ALAM manufacturers.

AMCMA, the independent's association, wilted, and many of its members, including the newly formed General Motors, which had grown out of the Cadillac operation, began applying for membership in ALAM.

Ford, however, still felt that he would eventually prevail, and counterattacked. His managers also placed ads—sometimes facing the Selden licensee ads—which they addressed: "To Dealers, Importers, Agents and Users of our Gasoline Automobiles," promising all of them protection against "any prosecution for alleged infringements of patents." Ford put up $12 million to back up his claim. The ads also spelled out Ford's reasoning, which was that the "Selden patent . . . does not cover any practicable machine, [and] no practicable machine can be made from it and never was."[18] At one point he called the patent a "freak among alleged inventions . . . worthless as a patent and worthless as a device."[19]

To help build up Ford's own reputation, the ads also contained some self-congratulatory text, which stretched the truth rather badly. They stated that "our Mr. Ford made the first Gasoline Automobile in Detroit and the third in the United States. His machine made in 1893 is still in use." Ford continued to claim that he had had a working auto in 1893, but the date was actually 1896. Nor was his the first gasoline-powered automobile in Detroit.[20]

The ads ran for months during the summer and fall of 1903, and created a very tense atmosphere in the automobile world. The ALAM manufacturers attached special labels on their cars to assure purchasers that they would be free of any lawsuits.

Selden himself, who had taken a backseat in many of these goings-on, now came forward and gave an interview to the Hartford *Courant*. In it, he insisted that any attempts to buck the licensing requirements were doomed to failure, and that his licensees had good and sufficient grounds for issuing injunctions. He implied that the patent had already been tested in the courts—which was not true. The courts had only ruled on the demurrers requested by the independents. Nevertheless, ALAM circulated the interview widely.

Although Ford was basically a shy man, he, too, was doing his bit to win the public over to his side—in some cases directly, in statements and appearances; often indirectly, in that the image and words of Henry Ford shortly came to dominate every piece of advertising put out by his forces. In each case, he was portrayed as a pioneer in the field, a winning racer, and a fighter for the rights of the public. Nor did the ads pull any punches: "It is perfectly safe to say that Mr. Selden has never advanced the automobile industry in a single particular . . . and that it would perhaps be further advanced . . . if he had never been born."[21]

In fact, Greenleaf states, "In all of the great industrial conflicts over patent rights in American history, nothing parallels Ford's shrewd instinct for marshalling popular feeling to his side."[22] It was this part of the battle that began to build his reputation as a monopoly buster in the world outside the automobile field. To buttress this reputation, the ads also stated that Ford had refused to become a member of the automobile trust. As we've seen, that's not quite the way it happened.

In any case, more than words were needed. Ford saw that as long as the Selden patent stood, it was not only his future that was at stake, but very likely, that of the entire automobile industry. Somehow, he had to beat that patent.

## *Fighting Back*

By this time, Ford had been thinking about and working in this field for a decade or more; he felt that his car and that of other early workers in the field did not owe anything to Selden's patent. At this point, the Selden licensees, noting that the Ford Motor Company was doing very well and seeing the the possibility of a real fight on their hands, put out feelers to Ford.

He happily turned his back on ALAM, hired new lawyers, and intensified his attacks. At the same time, the Selden interests brought suit against several more firms, including Ford's. The Ford suit turned into a monumental and drawn-out case. Heading up Ford's defense was R. A. Parker, a highly effective trial lawyer with long experience in this field.

Parker, it should be noted, had also led the Ford team in the earlier case, over which Judge Hough had presided. But Hough had had little patience with technical details, and Parker's careful explanations, both verbal and written, had tended to irritate him.

In the new case, Parker's talents had full reign. During the judicial proceedings, for example, Parker put the Selden side's expert witness under cross-examination for twenty-three days, with some seventeen hundred questions.

When Ford was examined, he continued to maintain that he had built and tested a car in 1892 and that his Quadricycle dated from 1893. Was this chicanery, or honest mistake? It's hard to say. But if the Quadricycle dated from 1893, it predated the issuance of Selden's patent. If 1896, it came after the patent.

Greenleaf argues, "It is a fair assumption that Ford deliberately advanced the date of his first car to indicate that his machine had performed successfully before Selden received the patent. The predating was ill-advised and unnecessary. It had no effect upon the outcome of the case, and it later created confusion in the recording of automotive history."[23]

For his part, Selden tried hard to argue that his engine did not grow out of the Brayton design, for that might very well have limited his patent to only that kind of engine.

The Selden lawyer, also highly experienced in this field, tried to broaden the engine argument, and argued that Selden had "appreciated the advantage of the compression cycle, adapted it to the purpose of a motor vehicle, showed one form of engine which was powerful in proportion to its weight, and thereby disclosed to the world for the first time an effective combination of a liquid hydrocarbon gas engine of the compression type with the other elements mentioned in his combination."[24]

When Selden himself took the stand, he was peppered with questions by Parker, but held up his end very well. He portrayed himself as a lonely inventor pursuing his idea in the face of all kinds of obstacles. He referred often to motor cars as his invention.

In a fascinating turnabout, Selden also accused Parker of asking unnecessary questions and thereby stretching out the trial so that Ford and the other independents could sell more vehicles before its conclusion.[25]

Through all this, the Ford Motor Company had continued to produce cars, lots of them. By early 1907, the company had outgrown the original Detroit plant and purchased a large tract in Highland Park, just outside Detroit, and began building the largest automobile manufacturing plant in the world. In doing so, Ford showed an amazing confidence in the rightness of his stand. After all, officially, the Selden patent was still in force. At the end of 1909, all the manufacturing operations were finally transferred to the new plant.

This was also an expensive plant to equip, for Ford was installing the most advanced machinery, and his staff had created new designs for fully half of that machinery. The typical automobile manufacturing facility of the time would turn out one cylinder at a time. Ford installed special machinery that would turn out fifteen cylinders at a time. Earlier hand methods of manufacture called for sawing, grinding, filing, and hammering in the assembly department. At the Ford plant, every piece was machined in a jig, to ensure interchangeability of parts. He also enticed some of the finest managerial talent in the country to come and play a part in this astounding enterprise. As part of the expansion, the company built assembly plants in other parts of the country as well, and then in other parts of the world.

Finally, on January 9, 1911, Henry Ford's courageous stand was vindicated. An appeals court overturned Judge Hough's 1909 decision. It stated clearly that the Selden patent was just too broad and that Ford therefore could not be said to be infringing on it. The terminology was so strong that the Selden forces saw little point to carrying the battle any further; and this time, it was ALAM that was disbanded.

## *Victory*

Thanks to Henry Ford, the American car industry was liberated from the tight restraints that Selden's group was trying to enforce. Ford's victory was widely reported, and his fame as a monopoly buster, a champion of free enterprise, rose accordingly.

At the same time, the company was really going to town. In 1913, it produced 168,304 cars, compared with 78,440 just a year earlier. The company was producing the famous Model T, and although the design was frozen, the company was continually refining the production process. In the following year, production again rose, to 248,307, an increase of over 80,000 cars; but employment *decreased* to 12,880, a reduction of almost 1,500 men.

Ford did not introduce the idea of mass production. He didn't invent interchangeability of parts or even the moving production line. But he combined these methods and brought the whole process to the point of creating a model for all the world to follow.

Even more important, however, was his effect on the social aspects of technology. Prior to Ford, the technical end of technology was moving right along; but the human end—the workers themselves—were being left behind. While some skilled workers benefited from the introduction of technology, he could see that in too many cases, the people working in the mills and factories were given very low wages for working long hours. Ford showed that with his methods, he could double the pay of the average worker while cutting his or her hours, and still make money. By making the worker a consumer, he created a whole new market for his own production. In addition, work on the farm was still isolating, boring, and often unprofitable. His low-priced vehicles changed that and had enormous consequences for anyone wishing to travel on his own.

In other words, Ford's obsession permitted him to carry out his plan, of which a very large part was his desire to produce lots of good, sturdy, reliable, but not fancy cars at a low price—a price that enabled the worker himself to become a purchaser of his company's automobiles. In those days, this was actually a revolutionary idea.

The Ford biographer and economist Anne Jardin describes Ford's philosophy as "a brilliant and daring synthesis of social needs with hitherto unobserved economic laws."[26]

## *Ford and the Patent System*

Ford was becoming so admired and famous that he entertained ideas of going into politics. "If I should go to the Senate," he stated, "one of the first things I would do would be to begin an agitation for the abolition of all patent laws. . . . [They] exploit the consumer, and place a heavy burden on productive industry. . . . [They are the] nemesis of free competition."[27]

We see how he felt about patents. He also acted on his beliefs. As the years passed, he, and then his successors at the Ford Motor Company, did not hesitate to apply for patents on any new developments they came up with—but only to be sure that others would not do what Selden tried to do, that is, to cut the ground out from under his competitors. But the company treats its own new developments, even the patented ones, as if they were common knowledge and freely grants permission to use these developments—without charge. Along the way, a trade association of auto manufacturers was formed, and all agreed that all such patents should go into a pool, to be used by all. Though Ford went along with the system in practice, he never officially joined with the other manufacturers or signed the agreement. The system hasn't been adhered to by everyone; there have been a few suits, but nothing of major proportions.

In the early 1940s, a Temporary National Economic Committee was brought together to determine whether restrictive practices were in use in the marketing phase of the automobile industry. It came to the conclusion that "the [cross-licensing] policy seems to have served the industry—and the public—well. There is no evidence that the progress of technology has not been as rapid as in kindred fields where grants have been fully exploited."[28]

## *Feet of Clay*

It was not all sweetness and light, and the giving was not all one way. Ford expected something from his workers as well. The advent of the moving production line gave him the option of speeding up the line. This he did in some cases, leading some to accuse him of exploitation rather than generosity. One woman wrote to Ford, claiming, "The chain system you have is a slave driver. . . . My husband has come home & thrown himself down & won't eat his supper—so done out! Can't it be remedied? . . . That $5 a day is a blessing—a bigger one than you know—but oh they earn it."[29]

Unhappily, Ford's later years were nowhere near as glorious as his earlier ones. His rapid rise to fame as an innovator apparently led him—and some members of the press—to think he had as much to offer in other areas, with the result that he did and said some very foolish things. In 1915, for example, he participated in and funded a so-called Peace Ship that traveled to Europe and sought to keep the United States out of World War I. Ford tried to convince Edison to participate, but Edison would have none of it.

Further, the project was so poorly organized, timed, and run that it ended up being an utter fiasco. The press turned against him and made him the butt of some powerful satire.

Curiously, the initial idea was presented to him by a Jewish woman, Rosika Schwimmer. This may have provided a spark that turned into an anti-Semitic flame. Ford later bought a newspaper that spewed the worst kind of anti-Semitic tripe for well over a year.

Even his business sense seemed to wither, and he continued to insist on one model in one color—the black Model T—long after its utility and predominance had waned. Once his "common man" began to get used to having his own inexpensive—and quite basic—car, it was no longer a thrill. And in the increasingly prosperous 1920s, he began to aspire to better, or at least more luxurious, cars. This Ford refused to supply. The result was a decline in his company's fortunes and in his reputation for infallibility. Finally, on May 31, 1927, Ford admitted his error, and production of the Model T stopped. Although the company had in the meantime produced more than 15 million of them, it had lost its primacy, and General Motors, which had a better feel for what the buying public wanted, had already bypassed it.

## *Legacy*

None of this, however, should detract from Ford's very real accomplishments. He helped establish a hospital and he created a major museum of early American ingenuity. But even these pale alongside his legacy in the automobile manufacturing field.

Thanks to his fighting ability and his faith in the competitive system, American automobile manufacturing blosssomed and has for many years been a major engine driving our industrial might. As of 2000, three of the four largest companies in the United States—General Motors, Ford, and Exxon Mobil—either made cars or supplied fuel for them. He created a huge middle class, that of well-paid industrial workers who could afford to buy the products their labor had produced.[30]

After him, the term *industrial* might also have included the workers, who for several generations at least could walk proudly and could afford to purchase the products their labor produced, including many luxuries once afforded only by the upper classes.

His influence was worldwide, at least in the more developed countries. The Ford factories that sprang up in many parts of what we now call the industrialized world served as models for other companies to emulate.

Further, he offered equal opportunity to blacks, paroled convicts, and the disabled; changed the face of the patent system; and in fact, created a new set of socioeconomic laws. Finally, and equally important, he set the standard for the cooperative use of patents and standards, which has served the world of American automobile manufacturing very well indeed.

As for the controversy itself, although ALAM was surely created with financial gain as its main objective, the mechanical branch of the organization did some very good work in technical standardization. But it was Ford who ensured that the field itself would be open to all comers, who could then profit from this standardization.

We can only wonder what we would be driving now had he not won his battle against Selden and company.

CHAPTER 6

# *Wright Brothers versus Curtiss, Chanute, Ader, Whitehead, and Others*

## The First Successful Flying Machine

Ask a group of youngsters, "Who invented the airplane?" Chances are they'll answer, "The Wright brothers." Are they to be commended? Or corrected? It will take us this entire chapter to decide.

For example, were the Wright brothers the first humans to fly in a heavier-than-air craft? No. Were they the first to fly in a powered aircraft? No.

Were they the first to build and fly such a craft in a controlled fashion? That's harder to answer. What is certain is that in the early part of the twentieth century, the Wrights patented a method of control that involved a coordinated change in the wings and rudder surfaces of flying machines.

Did they thereby have the right to argue that *anyone* who builds a craft with similar controls—no matter how different the configuration of the airfoil surfaces—was infringing on their patent and could be sued?

The situation sounds an awful lot like that faced by Ford in his battle against the Selden patent, doesn't it? The outcome, however, may surprise you.

### *Foundations*

As with so many other inventions, there was earlier work to build on, including powered lighter-than-air ships, as well as models and gliders. Among the more famous pioneers was the English baronet, Sir George Cayley, who began working on the problem in the early 1800s.

Remarkably foresighted, he set forth a clear definition of the problems involved in getting a powered aircraft to fly. He foresaw the final configuration as we know it today: a machine with enclosed fuselage, fixed wings, and tail surfaces, plus some sort of propulsion and control systems. After many experiments, including a series with rubber-band-powered models, he built the first successful man-carrying glider in 1853, which his manservant flew some five hundred yards across a valley.

Another tiller in this field was the American astronomer Samuel Pierpont Langley, who began looking into the problems of manned flight in 1886. The following year, he became Secretary of the Smithsonian Institution. With significant financial aid from the Smithsonian, he began his researches, first with models and then with powered craft. He came up with a design that had two wings in tandem—that is, one behind the other—and with a small steam engine and propellers between the wings. On May 6, 1896, one of his models, with a twelve-foot wingspread and powered by a small, one-horsepower steam engine, made two successful flights.

This inspired him to build a man-carrying craft, which he called the Aerodrome. Unhappily, it plopped ignominiously into the Potomac River shortly after launching on December 8, 1903—nine days before the Wrights' first successful flights. He tried again after the craft had been reconditioned, but the same thing happened. Langley tried to blame its poor performance on the launching apparatus, but the ridicule he faced from both the public and the press left him a totally disillusioned man. He died a few years later. Ironically, both Langley's Aerodrome and the Smithsonian were to come back and haunt theWrights later on.

The German experimenter Otto Lilienthal carried this work further. The owner of a factory that manufactured steam engines, he made some two thousand successful glider flights from 1891 until 1896, when he was killed during one of the flights—still seeking the secret of control. In his flights, he kept the glider under some sort of control by exaggerated movements of his legs and torso.

Octave Chanute, a French-born American civil engineer, felt there must be a better way of maintaining equilibrium than by thrashing about. He tried imparting stability to a man-carrying glider that he had built by spring-loading the wings, so that they might give way when hit by a wind and then spring back to continue the flight. That didn't work.

One of the men Chanute hired to help him was Augustus M. Herring. He, too, had tried to build in stability, but he worked with a movable tail. Although the idea of built-in stability turned out to be a flop, both Chanute and Herring were to figure strongly in the Wright brothers' future battles. Herring reappears periodically as a kind of slimy villain. The role of Chanute, as you'll see, is rather more complicated.

## *The Wright Brothers*

In spite of Lilienthal's lack of success, it seems to have been write-ups of his work that inspired the Wright brothers to become interested in manned flight. At the time, the bicycle was still an important mode of travel, and the brothers owned and operated a successful bike shop in Dayton, Ohio. This gave them the funds and the time to carry on their research program.

The brothers have often been portrayed as uneducated tinkerers, as mechanics who stumbled upon their solution by simple trial and error. Granted, their education was no more advanced than that of Edison, Ford, Stephenson, and others in this book. But their approach to the problem shows them to be far more than uneducated mechanics. Indeed, they embarked upon a surprisingly sophisticated program of research. In their campaign to find the solution, they moved along a carefully thought-out series of readings and experimentation.

Their first move, in fact, was a literature search, just as might be done today. They decided to contact the Smithsonian—which, after all, had sponsored Langley's research and experimentation and had published a number of articles and books on the subject. In the spring of 1899, Wilbur Wright wrote the Smithsonian and asked for a selection of their published materials on flight. Examples of the works the Wrights studied included a partial translation of an earlier book, *L'Empire de l'Air,* by a Frenchman, Louis-Pierre Mouillard. It was published in the Smithsonian's Annual Report for 1892.[1] Mouillard was fascinated by bird flight and tried to apply his observations to flying machines. He was not successful, but his excitement—indeed, his obsession with the possibility of flight by humans—seems to have rubbed off on the Wrights and may have provided some direction for them to start their observations.

There were also Chanute's *Progress in Flying Machines*[2] and Langley's *Experiments in Aerodynamics*.[3] Langley's book was a record of his experiments, plus a number of tables of engineering data derived from his work, which the Wrights used in designing their early kites and gliders. Three aeronautical annuals from the Smithsonian brought the subject right up to date.

At the same time, as with any scientist, observation formed an important part of the Wrights' program. They studied, and tried to tease out the secret of, the soaring and gliding of various birds. Of course, so had multitudes of other researchers, including many of the pioneers of flight. But it was the Wrights who provided the first solution to the puzzle. They noticed, for example, that when a pigeon wanted to turn, it did so in a way that no man-made craft up to that time had managed. Its smooth turns are banking turns, and it accomplishes them by changing the angles of its wings. To make a right turn, it will angle the left wing up and the right wing down; the left wing thereby generates more lift and will rise, while the reverse happens with the right wing.[4]

The Wright brothers never married and seemed to do everything together, though this doesn't mean they always agreed with each other. In Wilbur's biographical writings, for example, he insists that birds were indeed part of their instruction. Orville, who outlived Wilbur by thirty-six years, maintained in his later years that their observations of bird flight were more inspirational than instructional.[5]

However it happened, their recognition of the importance of banking turns was an important one. Until then, all human attempts at turns during flight, when they were made at all, were made by changing the angle of a vertical surface, the so-called rudder, as one would do in a boat. But the result of such a turn in the air, thanks to Newton's first law of motion, is a tendency of the craft to keep sliding in the same direction as the original direction of flight, and to sink as a result. The Wrights saw, apparently for the first time, that the more important control had to do with the attitude of the aircraft, that is, that a banking turn (with the inner wing dipping) would be necessary.

But how was this to be done with a fixed wing? The story goes that Wilbur was working in their bike shop one day and had just taken a new inner tube out of its box. While talking with the customer, he realized that he was holding the long, slender box at both ends and was idly twisting the ends in opposite directions. This was a eureka experience—and it led directly to their 1899 two-winged kite. Basically a pair of five-foot-long wings, the kite was guided by a pair of cords, each of which attached directly to sticks held one in each hand. By twisting the sticks in opposite directions, Wilbur (who took the first turn) found that he could indeed cause the kite's wings to twist in opposite directions, and that the kite's movements could be controlled just as they had hoped. The flight took place in an open field on the west side of Dayton, where the Wrights had flown conventional kites as boys.

The next step involved trying out the method in a manned glider. Because the glider could not be controlled by cords from the ground, it needed both vertical and horizontal control surfaces. They set up the all-important controls in such a way that the twisting motion of the wings, which they called wingwarping, would be coordinated with the proper turn of the rudder. This coordination of the wing surfaces with the rudder was to form the foundation for their basic patent.

## *Kitty Hawk*

But the eureka scenario was over. From now on, the discovery process needed experimentation, lots of it, which they carried out in the fall seasons of 1900 to 1902. The locale also had special requirements. They inquired of the weather service where in the United States they might find the steadiest and strongest winds. One of the locales suggested was the remote, storm-tossed Outer Banks along the northeast coast of North Carolina. So it was that Kitty Hawk, a small fishing village and site of a U.S. Weather Station, became the locus of much of their work from then on. The actual flights took place at an even more isolated area, Kill Devil Hills. A wide-open beach area about four miles from the village, it had neither trees nor hills for miles to the south and offered steady 10 to 20 mph winds, especially in September and October. By November, the weather usually turned too rough for safe operation.

## *Octave Chanute*

They realized that theirs was a large undertaking and that they would need some help. Wilbur wrote to Octave Chanute, a glider expert, on May 13, 1900, and asked a couple of questions. By then, this sophisticated Frenchman-turned-Chicagoan had already earned a significant reputation as a glider pioneer. We can imagine that he must have hesitated to get involved when he received a communication from the proprietor of a bicycle shop in little Dayton, Ohio. But something about Wilbur's letter caught his fancy, and he supplied answers. A year later, they invited him to participate in their project. Although he was already in his late sixties, he was happy to do so.

Getting to Kitty Hawk involved considerable difficulty—a long train ride from Dayton, then a thirty-plus-mile boat ride over rough water. It took Wilbur, who went out first in September 1900, almost a week to get there. When Orville arrived, they began their flights—constantly observing, adjusting, testing. What with periodic storms, wind gusts, mistakes, and crashes, and time for rebuilding the constant damages to their glider, the long hours of work translated into just minutes of airtime. But that short time suggested that they were on the right track. Then the weather said it was time to go home. In an attempt to gain some reputation in the field, Wilbur, the more facile writer, produced two articles on the aerodyanics of gliding flight and sent one each to a British and a German aeronautical organization (there being none in the United States), where they were published in July 1901.

They went back to Kitty Hawk that summer. Chanute came out in August and worked with them for a week. They tested their machine as both a kite and a glider. To the viewers who often came out from the weather station, the flights seemed impressive. But the Wrights were not satisfied. Neverthess, Chanute made arrangments for Wilbur to speak at the September meeting of the Western Society of Engineers on some of their work with gliders. He was very nervous, but did well. The highlight came when he showed a picture of the manned glider up in the air. It created a small sensation.

Then back to Kitty Hawk for more flights. In fact, these flights were especially disappointing—and confusing. There seemed to be something seriously wrong with much of the existing data—including data from Langley and Lilienthal—on which they were basing their designs. The Wrights decided they would have to depend on their own observations and calculations.

When they returned to Dayton, they carried out a new set of basic observations and calculations in the shop behind their bicycle store—angle of attack versus lift, airfoil configuration, and much more. To facilitate their measurements, they built a wind tunnel—a six-foot-long, sixteen-inch-wide box with a strong fan at one end and a window for viewing the parts being tested. To ensure the greatest accuracy possible, they mounted the fan separately and directed the air into the wind tunnel to prevent the fan's vibration from affecting the results.

The printed version of Wilbur's talk was reprinted in several important journals, including the Smithsonian Annual Report of 1902, and elsewhere overseas. Their reputation was growing fast.

Chanute included mention of their work in a chapter he was doing for a new book. As they became more knowledgeable, it seemed Chanute was becoming more the pupil than the teacher. He probably continued to be helpful, however, by acting as a kind of sounding board for their new ideas.

Then, in 1902, with a new and redesigned glider, based largely on their work during the winter, they carried out over seven hundred glider flights at Kitty Hawk. During that period, both Chanute and Herring had visited and spent ten days with the Wrights. Afterward, acting on information imparted by both these men, Langley sent a wire asking if there was time for him to come out for a visit. They turned him down, pointing out that it was late in the season and they were closing up shop for the winter. The real reason likely was that they saw Langley as a direct competitor and didn't want him to see their work close-up at this stage of the game.

Also, Langley, perhaps without thinking the matter through, had sent the Wrights a pamphlet he had recently authored, *The World's Greatest Flying*

*Creature*. In it he described the pterodactyl, an extinct flying reptile, and in comparing this creature with aviation activities to date, seemed to feel that his own craft was the only one worth mentioning. This did not sit well with the Wrights.

Their successful experiments led them to feel confident enough to warrant attempts at powered flight, but they could find no engine light enough and powerful enough to suit them. So they built their own, with the help of Charlie Taylor, a master mechanic whom they had hired in 1901 to take care of the bike shop while they were away. They spent the winter designing and building the powered model, a biplane glider with a seventeen-foot wingspread, and with, for the first time, two six-foot-tall vertical fins (later reduced to a single surface and called a rudder) to help the craft make turns.

The engine perched on the lower wing, and the two propellers were mounted at the back of the wings and faced backward. The pilot lay on his stomach to cut down on wind resistance. He used his hands to control the small horizontal surface at the front of the plane, while he controlled both the wingwarping and the rudder by pushing firmly with his left or right foot on a pivoted T-bar. In a later model, the wingwarp and rudder controls would be separated to give him greater (more sensitive) control.

In April 1903, Chanute had gone back to his native France and presented a report to the Aéro-Club de France on the progress of powered flight in the United States. His report was covered for the French magazine *La Locomotion* by Ernest Archdeacon, who included a rather curious statement, which he attributed to Chanute:

> Mr. Chanute, unlike most inventors who wish to keep for themselves alone the glory of their ideas, understood very well the necessity of helping one another in this tremendous task.
>
> Admitting that he was no longer very young, he took pains to train young, intelligent, and daring pupils, capable of carrying on his researches by multiplying his gliding experiments to infinity.
>
> Principal among them, certainly, is Mr. Wilbur Wright.[6]

In other words, the Wrights came off as disciples who were carrying out the plans and procedures of their mentor. It's possible that the idea was Archdeacon's. But it doesn't matter. What does matter is that the idea that the Wrights were not the real inventors of controlled flight was being planted, and would bedevil them for decades to come. Further, Chanute had given out some details of the wingwarping work, and this too turned out to be a major problem for the Wrights later on. Finally, in Chanute's extensive correspondence with the Wrights, and probably elsewhere, he also suggested that the discovery for which they were claiming credit was actually both ancient and well known.

The Wrights would recall Chanute's mentor/disciple statement with bitter resentment during their later patent battles. At the same time, Chanute himself would turn bitter over the Wrights' stratospheric success and what he saw as their selfish attempts to extort money from other aviation pioneers. At the moment, however, all was calm.

Finally, Archdeacon reported, Chanute closed his talk with the suggestion that the experimenters just were not ready to try motorized flight and would not be for another year or two. He quotes Chanute as saying, "[U]ntil then, it is useless, and even dangerous, to burden oneself with a motor, and I much prefer to use such an untroublesome and simple motor as gravity."[7]

Clearly, Chanute had little idea of how far along the Wrights really were. By September of that year, they were ready to test their new, powered glider, which they named *Flyer 1*.

## *Controlled Flight*

The upshot of all the Wrights' preparatory work was four short, but relatively controlled, flights in *Flyer 1* on December 17, 1903. The flights ranged from a 12-second/120-footer to one lasting just under a minute and covering 852 feet.

Curiously, only five people witnessed these historic flights. The Wrights had not invited the press. They preferred to conceal details of their discovery until they were properly protected by a patent, for which they had filed in March of that year. After the flights, they walked the four miles back to the village and sent a telegram to their father, who was primed to send it off to some of the newspapers. In other words, the brothers had no objection to an announcement of their accomplishment; they just wanted as few people as possible to get a close-up view of their invention.

Their prearranged notice announced the "world's first powered, sustained and controllable airplane flights." A momentous accomplishment, worthy of page one coverage? Far from it. The press release distributed by Bishop Wright received some placement in the papers, but little comment or emphasis. One of the problems was that the editors had heard such stories before, and the stories had always turned out to be worthless, wrong, or badly exaggerated. The Langley flights, for example, had occurred only nine days earlier. The flights had been touted as a major event and were well covered by the press. We've already heard about the results.

Some of the editors who received the Wrights' notice didn't even know the difference between an airship (balloon or dirigible) and an aircraft, and probably felt such flights had been done before. Finally, and perhaps most telling, the editors thought about who was making these claims: two tinkerers who owned a bicycle shop.

The following year, the Wrights did invite the press, but the fates were not kind. A combination of bad weather and bad luck led to a poor showing, and the Wrights lost some credibility. Shortly thereafter, however—on September 20, 1904—Wilbur made the first complete, controlled circle in a manned heavier-than-air craft. It doesn't sound like much, but it was the first in history.

Happily, this flight was witnessed by a reporter. He happened to be representing *Gleanings in Bee Culture,* but no matter. He had heard rumors of the goings-on at Kitty Hawk. Intrigued by the concept, he had made the two-hundred-mile journey by car—no mean feat in those days—and had happened upon this epic flight. He later sent a copy of his write-up to *Scientific American,* which simply ignored it.

There was another reason for the Wrights' mixed feelings about publicity: the patent process was not going as easily as they had hoped. Confident of the technical details, they had prepared the patent application on their own. But after they filed it in March 1903, the Patent Office rejected their application twice, claiming, among other things, that some of the statements were vague. The Wrights sought professional guidance and were advised not to file a patent on a powered flying machine—which was generally thought of in the same category as perpetual motion machines—but to concentrate on their method of lateral control, including both the wingwarping and simultaneous movement of the rudder. They worked further on the patent in 1904, and resubmitted their application. It was granted on May 22, 1906. Word of their successes had been spreading and was no doubt a major aid in the granting of the patent.

Meanwhile, Chanute, with whom they still maintained cordial relations (the damage done by his talk was not to surface until later), was urging them to start flying in public, so that the world could see what they had accomplished. But they knew full well that once the method became widely known, there was the great likelihood that many would-be designers and aviators would be tempted to use their method without proper licensing. They reasoned that it would be better to hold the secret close to their chests until they could get some sort of major contract with the U.S. Army, or some other major buyer. This would provide the funds for patent litigation, which they expected (correctly) would be expensive.

But the government was no easier to convince than was the press, and the first contract, with the U.S. Signal Corps, was not signed until February 1908. At about the same time, they signed with a French company, which was to manufacture their improved model. They used the time until then to improve their product and came up with the Wright *Type A Flyer.*

Both contracts called for demonstration flights. Neither brother had flown for quite a while, and the control system was still very primitive. On the other hand, it was far in advance of anything else around, and when Wilbur went to France he put on demonstration flights that seemed spectacular at the time.

In the meantime, Chanute's 1903 talk and the subsequent publication of it had caused an explosion of interest in flying, and Europe, especially France, saw some solid results.

Many of the planes being built, tested, and flown used some primitive form of control and were managing longer flights. By 1906, one plane won a prize for a flight of a couple of hundred feet. On July 25, 1909, Louis Bleriot, a Frenchman, made world headlines by flying across the English Channel. His craft was a monoplane with an adaptation of the Wrights' wingwarping mechanism.

But the flying bug had also infected the United States, and it was an American, a bicycle manufacturer and motorcycle racer named Glenn Hammond Curtiss, who was to create the biggest headache for the Wrights.

## *Glenn Curtiss*

When not winning motorcycle races in the early years of the twentieth century, Curtiss was building and supplying lightweight engines for use in dirigibles.[8] His engines were good enough to attract the attention of others interested in powered flight, including Alexander Graham Bell, who had turned his attention from the telephone to aircraft. Curtiss, Bell, and a few others formed a group, the Aerial Experiment Association, in 1907 and began work on powered aircraft, basing their designs on work that had been done by the Wrights and by the French team of Gabriel and Charles Voisin.

Curtiss, ironically, had been hesitant to join this impressive group, which included not only college-trained people but also a well-known scientist. Curtiss's own education, like that of the Wrights, had been very limited. Yet his contributions to the group were considerable. He also went on to develop a successful amphibian aircraft and to found his own airplane and engine company.

Although the association was well aware of the wingwarping principle, the members tried in every way to go around it, partly for fear of infringing the Wright patent, and partly because they felt that wingwarping might put too much strain on the wings. Their first craft, in fact, had no capability of lateral control at all. But then the group came up with an alternative—hinged, movable surfaces built into the wing tips (later called ailerons) that performed the same function as the Wrights' wingwarping and that they hoped would get around the Wright patent.

By the summer of 1908, they were on their third plane, called the *June Bug,* and on July 4, Curtiss piloted it on a successful flight of over a mile before several hundred people. It was in truth the first preannounced, controlled, successful, public flight in the United States.[9] He even beat out Orville, who, with a craft that was much improved over the Wrights' early *Flyers,* didn't get to make his first public flights until September. Earlier, in July, Curtiss had also won the *Scientific American* trophy for managing to fly a one-kilometer course without crashing.

The *June Bug* was, in several ways, a much more modern craft than even the latest Wright *Flyers.* The latter still used skids rather than wheels; they had to be sent up by a complicated catapult arrangement; they used a basic automobile engine for power; they used two vertical sticks for control; and mainly, they still depended on wingwarping for lateral control. The *June Bug,* in contrast, featured tricyle wheels for takeoff and landing, a control wheel, and a sophisticated, lightweight, eight-cylinder engine. Most important, it used ailerons for lateral control. These remain the basic lateral control mechanism on all modern aircraft, and in the modern version, provide a far easier and surer method of control than the wingwarping.

In fact, a group of determined revisionist historians insist that it is Curtiss who should properly be given the honor of having flown the first truly controllable aircraft. The problem is that *control* is a relative term. Jack Anderson, one of the revisionists, maintains that the wingwarp-controlled Wright aircraft were never really controllable in any modern sense and were very prone to stalling. He also quotes Grover Loening, who worked for the Wrights in 1913, as saying that flying in an early Wright aircraft was "like sitting atop an inverted pendulum ready to fall off on either side at any moment."[10]

Another writer, Charles Grey, describes the Wright *Flyers* as a dead-end design.[11] The problem was that all the craft of the time were hard to fly and were dangerous. They continued to be built and flown because flying was new and exciting and no one knew any better. The first recorded death occurred in September 1908, when an Army lieutenant, Thomas Selfridge, went up as a passenger in a Wright *Flyer* with Orville at the controls. The craft got up to about 150 feet in altitude when propeller problems caused the craft to go into an uncontrolled dive and crash badly. Selfridge was killed, and Orville was severely injured.

In the years 1909 to 1913, nine of fourteen army officers were killed taking flight training. Most of them were flying Wright planes, but several Curtiss craft were involved as well.[12]

## *Wright v. Curtiss*

The Wrights, stung by the public flight of the *June Bug* in 1908, decided it was time to make a move. Their feeling was that any form of movable wing section was at the very least an offshoot of their wingwarping idea and so was covered by their patent, which had already been granted. Though they had until then freely permitted use of their idea for research purposes, they felt they had the right to ask for fees from anyone who stood to make money from their idea, generally for manufacturing aircraft but also for racing or exhibiting.

Curtiss certainly fell into all of those categories. He had visited the Wrights at their bicycle shop in 1906 and 1907, and the Wrights felt strongly that his rapid development in the aircraft field came from information he apparently gleaned during those visits.

Although the Wrights became entangled in several major legal battles, both in the United States and abroad, the case involving Curtiss was by far the most bitter and went on for years.

The opening salvo came in a letter from Orville to Curtiss dated July 20, 1908: "I learn from the *Scientific American* that your June Bug has movable surfaces at the tips of the wings. . . . We would be glad to take up the matter of a license."[13]

But Curtiss had no intention of paying a licensing fee to the Wrights. He felt that the Aero group's ailerons would not fall under the Wrights' patent.

There was another reason. The devious Augustus Herring had talked his way into a partnership with Curtiss, and they formed the Herring-Curtiss Company. It turned out to be a big mistake for Curtiss, and extricating himself from the partnership was expensive.

In August 1909—after the Wrights' demonstration flights for the U.S. Signal Corps were finally completed—they brought suit against both the Herring-Curtiss Company and Curtiss himself for infringement of their patent. Officially titled *Wright Company v. Herring-Curtiss Company and Glenn H. Curtiss*, the real target was clearly Curtiss, and the suit came to be widely known as *Wright v. Curtiss*.

Trying to counter the publicity generated by the suit, Monroe Wheeler, the president of the Herring-Curtiss Company, issued a statement claiming that the company was in possession of patents taken out by Herring that predated the Wrights' claims. He based this on nothing more than Herring's misleading statements and apparently never asked to see the patents. Herring's claims fell apart shortly, but the false claims continued to circulate among the press and public.

Chanute lost no time in taking Curtiss's side. When asked by the press for his feelings, he stated his belief that their wingwarping idea was a concept that had been played with by many, including Leonardo da Vinci. Another article reiterated Chanute's feelings that his own contributions were not being properly recognized by the Wrights. In the article, it came out with the usual reportorial beefing up. The reporter said of the Wrights: "Their persistent failure to acknowledge their monumental indebtedness to the man who gave them priceless assistance has been one of the most puzzling mysteries in their careers."[14] They were indeed indebted to him, but mainly for his encouragement and support in the early years.

A short time later, however, their long friendship began to reassert itself, and Chanute wanted to make amends. But he was in bad health, and was not up to it. Chanute died November 23, 1910.

In the meantime, the patent battle was heating up—and took some fascinating turns. From the time of the Wrights' successful filing in 1904, their patent attorney, Harry Toulmin, filed also in all the European countries that seemed likely to develop capability in aeronautics. But now Chanute's 1903 report to the Aéro-Club de France came back to haunt the Wrights.

Although it's not clear whether Chanute's report was actually put to use by European aircraft pioneers, he apparently had revealed enough for it to constitute "prior disclosure." The result was that the Wrights' first filing date protected them in the United States, but not in many of the European countries where they had filed.

In 1908, the same year in which Curtiss was making a name for himself in the aviation field, the Wrights, too, were building their reputations—Wilbur in the United States and Orville in Europe. As the Wrights' flights and successes became better known, and as some of the patent implications became clearer, others awoke and began to claim priority. Sometimes questions of priority have to do with pride. Not here. If someone else could prove his priority, the Wrights' patent would no longer be valid.

The list of those who claimed to have flown before the Wrights, or of those who made claims for them, is very long. Among the best known are Langley, Herring, Clément Ader, and Gustave Whitehead.

Herring, for example, continued to claim priority. Several accomplices pushed the idea that Herring had done much of the work for which the Wrights were getting the credit.

Ader, a respected French engineer and inventor, had designed a bat-like, steam-driven contrivance, which he later claimed had flown almost a thousand feet, and that it had been done before the Wrights' flights and patents. His claims were later shown—to the satisfaction of most aviation historians—to be untrue, but not before they were used against the Wrights.

In France, where so much of the original upgrading was done, the idea that priority should go to the United States was, and apparently remains, an unpleasant thought. Claims continue to circulate that Ader had indeed gotten his craft up off the ground in 1890 and again in 1897, both well before the Wright's flights. The French, in fact, have continued to honor his name as "the man who gave wings to the world," for example, at public ceremonies in 1950 and 1990. But solid proof is hard to come by. And even if the craft did rise, is that flying?

Whatever the real results, one of Ader's machines *(Avion III)* was exhibited at a museum and then at a major aeronautical exhibition, along with the claim that it had flown in 1897, six years before the Wrights' historic flights. The well-known French brothers, Charles and Gabriel Voisin, not only backed up Ader's claim but also maintained that the Wrights had learned from Chanute, who in turn had learned what he knew from Ader.

These claims were among the many factors that had to be taken into account when the patent battle with Curtiss went to court. Curtiss was represented by Emerson Newell, who showed early the tack they were going to take by referring to the Wrights as "airplane chauffeurs."[15]

Ironically, the suit was being tried in front of none other than Judge John R. Hazel, the very same judge who had presided over the Ford-Selden case. Not surprisingly, he applied the same sort of broad reading of the patent that he had taken on the Selden patent. But in this case, it worked to the benefit of the Wrights.

He handed down his opinion on January 3, 1910. The Curtiss ailerons, he argued, were covered by the Wrights' wingwarping claims. He granted the Wrights a temporary injunction that would prevent the defendants from producing aircraft or making exhibition flights for profit.

Curtiss appealed and was granted a temporary stay of the injunction upon posting a $10,000 bond. Less than six months later, his appeal was denied and his bond was returned to him. The appeals court argued that the defendants' control method was not different enough to warrant an injunction. The original decision against Curtiss stood.

In April 1910, the Aero Club of America, which sponsored many races and exhibitions, agreed to pay the Wrights 10 percent of the gate receipts.

But the fight was not over, not by any means.

## *Twists and Turns*

As the battle dragged on, the same sort of feeling that had led to adulation of Ford now began to work against the Wrights. Curtiss's side argued that the Wrights were trying to set up, and to profit hugely from, a monopoly situation. The Wrights' side pointed out that the brothers had given explicit permission for their invention to be used without charge by anyone doing experimental work in the field of aeronautics and that it seemed only fair that they should not let others profit from the years of work and funds that they had invested—and the dangers they had faced. Orville, for example, was involved in five serious accidents during tests of their aircraft. The last one, in 1908, inflicted injuries that were to cause him severe pain off and on throughout the rest of his life.

Ford had argued that a broad interpretation of the Selden patent would place a noose around the fledgling automotive field. And he had won on that basis. The Curtiss side argued that the same thing would happen in the world of aeronautics.

The arguments went on for years, involving the two sides, their lawyers, the courts, the press, even the public. Fred Howard, who did a major biography of the Wrights in 1987, says the seven years of litigation left "a sordid trail of hatred, invective, and lies that muddy the pages of aeronautical history to this day."[16]

The Wrights, backed by an East Coast investor group that had bought the Wright patents, could afford the huge costs; the Curtiss group had more difficulty, particularly when the costs of disengaging from Herring began to mount. But that didn't stop them from pulling out all stops. Experts were brought in, by both sides, who helped argue many basic questions. Is the Wrights' patent too broad? Were they entitled to profit from others' work in this field? Was their invention not an invention at all, but a simple adaptation of a phenomenon that was common in nature?

How much did their work depend on the discoveries made by others? Mouillard, for example, whose book on bird flight had inspired the Wrights, did have a patent on wingwarping. Examination of the patent, however, showed that it had nothing to do with lateral control, nor did it present a method of attaining such control. The arguments went on and on.

During the fracas, both Curtiss and the Wrights were doing a lot of exhibition work. At least Orville was; Wilbur was kept busy with the patent suits they had brought in both the United States and Europe. Many decisions had to be made. They had, for example, obtained an injunction against Claude Grahame-White, a popular English sportsman and flyer, prohibiting him from flying races and exhibitions in the United States. But they also decided that they would not bring charges against racers from the United States or abroad at an international meet being sponsored in 1912 by the Aero Club of America.

At the same time, Wilbur was still deeply involved in the patent battle with Curtiss and was complaining that the time eaten up by the constant hearings and postponements already had destroyed three-quarters of the value of their patent. But the battle may have taken its toll in another way. For, Orville complained later, Wilbur was both stressed and distressed by the court battle, which may have affected his resistance to disease. In any case, he came down with a fatal case of typhoid fever. Orville at age twenty-five had successfully won a bout with the disease. Wilbur at age forty-five was not so lucky. He succumbed on May 29, 1912.

## *Finally?*

On February 27, 1913, Judge Hazel confirmed his earlier (1910) decision, and Orville got another restraining order against Curtiss. Curtiss again appealed. A year later, on January 13, 1914, the Court of Appeals decided against Curtiss. This seemed, finally, to end the legal proceedings, with a clear victory for the Wrights.

Orville decided that he would, in general, take it easy on American manufacturers who had bypassed the patent in the past—except for Curtiss. At the beginning of the patent battle, the Wrights and Curtiss had maintained a public appearance of cordiality. As it proceeded, the air began heating up; but Orville reined himself in until the patent situation was finally decided. It was now time, he felt, to vent his real feelings. In a *New York Times* interview dated February 27, 1914, Orville's position was clear: "For all other aeroplane manufacturers except Mr. Curtiss . . . he would adopt a policy of leniency."[17]

He also stated his belief that Curtiss had borrowed generously from him and his brother: "We met Mr. Curtiss [in 1906] and told him—then an incredulous unbeliever—about our flights. . . . We told him all there was to tell."[18]

Orville even blamed Wilbur's death on Curtiss, arguing that the long legal battle had "worried Wilbur, first into a state of chronic nervousness, and then into a physical fatigue, which made him an easy prey for the attack of typhoid which caused his death."[19]

Curtiss responded with a press release that stated, "I never had an item of information from either of the Wrights that helped me in designing or constructing my machines or that I ever consciously used." More generally, He called called Orville's charges "absurd, if not malicious."[20]

Was this the end? No. Henry Ford, still a firm opponent of strong patent protection, here entered the scene. He urged Curtiss to battle on and suggested that he obtain the services of W. Quentin Crisp, who had helped Ford break the Selden patent. Crisp came up with the idea of rebuilding Langley's Aerodrome. If they could show that that it was actually airworthy, that the problem had indeed been in the launching apparatus, then the flights would predate those of the Wrights, and this might convince the court that the Wright brothers should not be granted the broad protection it had given them.

To help their cause, they enlisted Albert Zahm, who, as the director of the Langley Aeronautical Laboratory at the Smithsonian, was the custodian of the Aerodrome remains. With his help, Curtiss was given $2,000 (in 1914 dollars) to rebuild Langley's craft. But in a clever ploy, they made a variety of small improvements—in the same way as the upgraded Selden vehicle was supposed to have shown its inherent abilities. These included a better propeller, a stronger engine, and a set of the Bell group's ailerons. If they could get the craft to fly, it might be taken to show that the Aerodrome's inherent design had indeed been flightworthy.

The ploy didn't work for several reasons. The main one is that although the craft did rise, it only managed to stay up for five seconds. Its wings broke apart before it could be tested for turning ability. The courts were not impressed.

Further, the attempt did nothing to endear Curtiss to Orville. Still, both sides claimed victory, and Curtiss got some revenge in another way. The Smithsonian next had the Langley craft rebuilt again, but this time to the original design. Starting in 1918, the craft was displayed for years, along with a caption giving it credit for being "the first man-carrying aeroplane in the history of the world." All of this was also reported in the institution's annual reports, where it became part of aviation history. Curtiss must have chuckled at that.

Orville, of course, was furious. He still had the remains of *Flyer 1,* the craft that had made those historic flights. Surely the appropriate place for a reconstructed *Flyer 1* would be at the Smithsonian. But he certainly was not going to share credit with Langley for the first controlled, manned flight, or even appear to be a runner-up. As a result, *Flyer 1* ended up on display at the Science Museum in London, where it remained on display for twenty years.

## *Forced Agreement*

Henry Ford, only a few years earlier, had helped automobile manufacturers break free of the Selden patent, permitting the American automotive industry to flourish. Curtiss, in spite of Herculean efforts, was not as successful in the world of aircraft. Curtiss defenders argue that the result was a stultification of the American aircraft industry.

Phil Scott, who edited a collection of writings by aviation pioneers, puts it this way: "While aviation continued its breakneck development in Europe, in America it was stymied by the withering grasp of the Wrights' patent suit."[21] One result was the clear superiority of European aircraft during World War I.

Curtiss himself, as well as a number of his followers, felt that his legal activities helped keep the field from freezing up entirely and that the fledgling aviation industry therefore owed him a solid debt. As he wrote to a friend, W. I. Chambers,

> [T]he litigation has cost us forty or fifty thousand dollars and the industry—or what there is of it in America—is entirely the result of our taking up this case as we have, single-handed, and fighting it out. Had we not taken this stand, the Wright Co. would have been in a position to enjoin all manufacturers and the whole industry would have been monopolized. . . . In fact, any machine which could fly, including hydroplanes [an important development by Curtiss] and all would come under the jurisdiction of the Wright patent.[22]

By 1915, most of the patent suits had been settled in one way or another, and the Curtiss Airplane Company was on its way to becoming the largest aircraft manufacturer in the United States. Ironically, Curtiss by then had several important patents of his own under his belt—including that for ailerons—and he now was notifying other aircraft manufacturers that they must obtain licenses from him![23]

According to the revisionist Ray Ettington, "Given current U.S. Patent Office practice it is doubtful that the Wrights' patent would have been upheld today." Ettington, who is also a docent at the Glenn H. Curtiss Museum in Hammondsport, New York, adds that present regulations include the comment that "the laws of nature, physical phenomena and abstract ideas are not patentable subject matter."[24] But, the revisionists contend, that's just the sort of thing the Wrights had succeeded in patenting.

Their patent situation has been compared with Selden's tying up of the automobile market, but I feel there's an important difference. The Wrights actually followed through and produced a workable solution to a major problem. So, clearly, they were entitled to some sort of patent protection. Whether they were entitled to as broad a cover as they received is a different question.

C. S. Roseberry writes in his biography of Curtiss, "There were no guideposts for either case [Selden or the Wrights], and the maladministration of the patent laws was notorious. . . . Incidentally, the 'broad construction' viewpoint relieved judges of the necessity of dealing with intricate technical details which baffled understanding."[25]

As the litigation dragged on, Curtiss tried to work out some sort of agreement, and in 1915 he and Orville even held a brief meeting, but Orville was adamant and it went nowhere. By then, however, World War I was in full swing, and the United States government decided that patent protection was hampering operations of many companies.

With the advent of the war, however, the U.S. government imposed its will on the patent situation, and decided that all pertinent patents should go into a patent pool. These would be shared by all aircraft manufacturers as an aid to the war effort, with a flat fee charged for their use. The Wright Company (by then Wright-Martin) and the Curtiss Aeroplane and Motor Company each received $2 million as payment for all patent rights.

With the cross-licensing arrangement established in August 1917, *Wright v. Curtiss* was finally rendered moot—which was all right with Orville. With Wilbur's death, he seemed to have lost his enthusiasm for aviation. By 1918, he had stopped flying altogether, though he did continue doing some research in the field. Among his inventions were an automatic pilot system and a split-flap airfoil. The latter device found use on some American dive-bombers during World War II.

On July 5, 1929, twelve of the Wright- and Curtiss-affiliated companies were merged into the Curtiss-Wright Corporation, which still operates today.

## *The Smithsonian Institution Again*

Orville would die of a heart attack at the age of seventy-six on January 30, 1948. Which means he would live long enough to go through yet another unhappy experience with the Smithsonian Institution. After Albert Zahm had played his part in the Smithsonian/Aerodrome fracas, he left and became chief engineer at the Curtiss Airplane Company. In 1929, he came back to the Smithsonian, this time to the recently established Guggenheim Chair of Aeronautics—which of course put him in a good position to carry on his campaign against the Wright name.

For example, he began to champion the name of Gustave Weisskopf. After Weisskopf emigrated here from his native Germany, he had changed his name to Whitehead, which is how he is known in the United States. Whitehead, and then several supporters, made a number of unsupported claims of his having flown before the Wrights. Although he eventually gave up aviation for religious fanaticism, his claims were, unfortunately, picked up by several publications and presented as fact rather than dreams of glory. In 1937, Zahm arranged for publication of a book that described *The Lost Flights of Gustave Whitehead.* It was written by Stella Randolph, who had co-authored an earlier, equally imaginative article about Whitehead.

In 1945, Zahm came up with his own entry, *Early Powerplane Fathers.* In it, he covered the work of four men whose work purportedly preceded that of the Wrights. Its star figure was none other than Gustave Whitehead. A network radio program picked up on the idea, stating that no one had ever proved that Whitehead had *not* made such flights. Then the *Reader's Digest* picked up the idea and ran with it.

All of this explains why the remains of the Wright *Flyer 1* ended up in the Science Museum in London and stayed there for twenty years. Zahn finally retired in 1945, and his successor, Charles Abbot, tried to straighten out this ridiculous situation. He approached Orville and asked about bringing *Flyer 1* home. Orville was interested, but would agree to its display only if the Smithsonian would agree that under no conditions would it or its successors "publish or permit to be displayed a statement or label in connection with or in respect of any aircraft model or design of earlier date than the Wright Airplane of 1903."[26]

*Flyer 1* was finally returned to the United States and put on permanent exhibition in December 1948—eleven months after Orville's death at age seventy-six. It remains there today, in the prominent position it deserves.

This leads to yet another hard-to-believe outcome. The Smithsonian's earlier backing of the Whitehead claims gave rise to a strong advocacy group that continues to operate today. An Internet search for Gustave Whitehead brought up hundreds of sites, including mention of a Gustave Weisskopf Museum in Leutershausen, Germany. In several of the sites, a most ironic claim is made—that a major reason Whitehead is not better known today is the agreement noted just above. One site reports that "Connecticut State Senator George L. Gunther . . . has a considerable collection of Whitehead material and continually pesters the Smithsonian to release themselves from the Wright Brothers' contract so they can research the claims of other flights prior to the Wrights."[27]

In recognition of their contributions to the world of aviation, both of the Wright brothers were elected into the Hall of Fame for Great Americans at

New York University—Wilbur in 1955 and Orville in 1965. Glenn Curtiss, embittered by the drawn-out patent war, had left the field early to become a real estate developer in Florida. Still, in recognition of his contributions, he was given the Congressional Medal of Honor in 1930. He died the same year, at age fifty-two. As mentioned, it was the ailerons developed by him and the Aerial Experiment Association that have become the standard in today's aircraft. By 1915, even the Wright Company was using ailerons on its updated designs.

Wilbur and Orville were elected into the National Aviation Hall of Fame in 1962, Chanute in 1963, and Curtiss in 1964.

CHAPTER 7

# *Sarnoff versus Farnsworth*

## The Fathers of Television

One of the great, defining images of the twentieth century is that of Neil A. Armstrong setting foot on the moon on July 20, 1969. A manned landing on the moon! It was a miracle of engineering.

Armstrong's moonwalk was seen by hundreds of millions of people at the very moment it was happening. That, too, was a miracle of engineering. And the time and effort put into the development of television surely matches, and perhaps exceeds, that of the moon flight. Certainly, television has had far more impact on people's lives.

Yet, ask anyone, "Who invented television?" and you're likely to draw a blank stare. Or your respondent may offer, tentatively, "David Sarnoff?"

Where, and who, are the inventors of television?

Thirty years before the moon walk, and for a cosmic instant, the name Philo T. Farnsworth was poised to stand alongside Edison and Westinghouse as one of the foremost inventors of our time.

As in all cases of science and invention, there were others who deserved credit for their contributions to the world of television. But it was Farnsworth who first came up with the idea for a totally electronic television system in 1921, at the age of fifteen! He filed the first patent for an all-electric system in 1927. In 1937, at the age of thirty-one, he was rated one of the ten greatest living mathematicians.[1] By 1939, he was considered one of America's Top Ten Young Men.[2]

After years of fighting him, the corporate giant RCA had agreed to pay him a million dollars for the use of his patents—the first and only time in its history that it had to do so. A new organization, Farnsworth Television and Radio Corporation, had just been organized in 1938, with Farnsworth as vice president in charge of research. The company had just purchased and was outfitting two major plants in Indiana to turn out radio and television sets. Farnsworth had just signed a major cross-licensing agreement with AT&T.

Yet, by the end of that same year, it was David Sarnoff whose name became associated with television in the pages of the press and the mind of the public. History quickly began to forget Farnsworth. By 1944, the Television Broadcasters Association saw fit to bestow the title of "Father of American Television" on Sarnoff.[3] In 1950, the Radio and Television Manufacturers Association seconded the motion.[4]

If ever there was a David and Goliath story in the history of technology, the battle between Farnsworth and Sarnoff was it. But this one did not have a Hollywood ending. By 1957, Farnsworth could appear on the television show, *I've Got a Secret,* and stump a panel of experts. He stated, "I invented electronic television. Who am I?" Not one of the panelists knew of him.

## *Early Years of Television*

By the turn of the twentieth century, the Italian physicist and inventor Guglielmo Marconi had shown that telegraph signals could be sent over air waves. The broadcasting of information—as differentiated from the point-to-point communication used in the telegraph and telephone—suddenly made sense. Radio, also called wireless, came into being.

But if speech and music could be sent, why not pictures? It seemed a reasonable follow-up, but there's one major difference. The telegraph, telephone, and radio have something important in common: they all involve serial transmission of information. That is, the dots and dashes, and the sounds you emit as you speak, come one after the other and can be transmitted as such. The same holds for notes in a musical composition. This simplifies the communication process.

Pictures are different. A picture is a two-dimensional entity. It can't be sent via radio waves until a very important process takes place first. The image must be broken up into tiny parts, each of which must somehow be transmitted, one after the other, and then it must be reconstituted at the receiving end. The idea is simple; the technology is not. Each tiny light element must be converted to an electrical signal, transmitted—by wire or airwave—to the destination, converted back to light, and then reassembled at the end. If the process can be done quickly enough, the eye sees the collection of elements as a complete picture.

In 1894, Paul Gottfried Nipkow, a Russian engineering student, took an important step along the way. He came up with the idea of a rapidly rotating disk, 12 to 14 inches in diameter, with a series of some 25 to 50 holes arranged in a spiral near the rim. The disk faced a lens system on one side that directed light onto it from the scene being viewed. As the disk spun, the holes dissected the light into tiny bundles which went through the disk holes and fell onto a photoreceptor. The photoreceptor in turn emitted a series of electrical signals that reflected the strength of the successive light signals passing through each specific hole.

These signals controlled an associated light output at the receiving end. Because this light was placed behind a second disk that was rotating at the same rate as the first one, the appropriate signal would shine through the appropriate hole in the second disk.

Thanks to a biological phenomenon called persistence of vision, as long as the disks spun at a speed of at least sixteen rotations per second, the viewer saw, in the plane of the receiving apparatus, a reproduction of the original scene. Although the Nipkow disk produced a very coarse representation, it was a way to accomplish the job, and the basic idea for this mechanical system was picked up and improved by a number of later inventors. Improvements included both speeding up and enlarging the disk. It wasn't much, but it was the beginnings of television.

## *Sarnoff*

The year 1923 was significant in the history of television, although one would hardly have known it at the time, for commercial television was still decades away. The twenties was the era of the radio, with many manufacturers producing both parts and sets.

RCA (Radio Corporation of America), which had been formed only a few years earlier, had already managed to corral all the important patents covering this still new and thriving technology—by buying up the patents or the companies holding them. This included those of the all-important vacuum tube, the predecessor of the transistor.

But, as had happened earlier in automobile manufacturing, many of the smaller manufacturers simply ignored the licensing requirements. David Sarnoff, RCA's commercial manager, was a smart, tough businessman; he chose not to swallow these losses and decided to go after the offenders. He tightened up the requirements and set up standards that determined who could buy parts and who could not. Even to the favored manufacturers, he would only sell the all-important tubes with the sets themselves. If a distributor claimed he was replacing burned-out tubes, he would have to return the old tubes first.[5]

Sarnoff was not alone in this. RCA was part of a larger radio trust, which included General Electric, Westinghouse Electric, and AT&T, and all were facing charges of monopolizing the radio manufacturing industry and conspiring to restrain competition. The government instituted antitrust proceedings in 1923. After much cogitation and negotiation, and to prevent further damage, General Electric, Westinghouse, and AT&T surrendered their interests in RCA.

RCA therefore emerged as an independent entity, which was just fine with Sarnoff, who began moving up in the hierarchy. The company prospered and would be instrumental in the building of Rockefeller Center in Manhattan, with the RCA building as one of its major edifices.

Although the patent pool remained intact and was basically under RCA's control, the company agreed to license out the patents to anyone, for a fee to be sure, but without restriction. But Sarnoff was relentless, and competing businesses that failed to pay the required licensing fees were taken to task. RCA was large enough, and powerful enough, to buy companies that tried to stand in its way. Alternatively, lawsuits and countersuits were a common occurrence in business dealings.

Always looking toward the future, Sarnoff saw the possibilities offered by the technological infant already called television. He made up his mind to do in television what he had already accomplished in radio, and he began tracking its progress very carefully.

He saw that John Logie Baird in England and Charles Francis Jenkins in the United States were both putting together television systems, which they were hoping to market. Baird was using the mechanical Nipkow disk to create the picture elements, while Jenkins employed a variation on the idea. Their receiving devices utilized the already widely available cathode-ray tube—a larger tube that had its end surface covered with a chemical substance that would emit light when struck by electrons. Although the results still left much to be desired, the idea was exciting, and demonstrations wowed viewers. Looking ahead, Sarnoff wondered if there was some way to dispense with all the mechanical equipment in the transmitters; that is, to do the whole job electronically.

In the meantime, Farnsworth was moving along, slowly, and by 1927, he had managed to demonstrate a basic system that could transmit the image of a line. No more than a line at first, but this progressed to a triangle, a dollar sign, and by 1929, photographic images. He was on the right track.

## *Zworykin*

Vladimir K. Zworykin, a Russian emigré employed by Westinghouse Electric, had a similar idea. In his younger days in Russia, he had studied with Boris Rosing, a professor of physics at the St. Petersburg Institute of Technology. Starting around 1910, Rosing invited him to participate in a project that he called electrical telescopy (seeing at a distance). Rosing had some good ideas, and did some pioneering work in electronic scanning, but the equipment available at the time just was not up to the task.

After a series of adventures having to do with World War I and its aftermath, Zworykin emigrated to the United States, and in 1920, found employment with Westinghouse Electric, which was still part of the radio trust at the time.

Still interested in the idea of television, Zworykin designed a system, which he described as an electric scanning system, and with Westinghouse's help, he filed for a patent in 1923. What he had to show, however, was not very impressive; Westinghouse's management, much more interested in their radios and their refrigerators, paid little attention to Zworykin's television work and wanted him to continue work in other areas. Frustrated, he left to work in another, smaller firm; but he got nowhere there, either, and he returned to Westinghouse.

Baird and Jenkins continued to demonstrate their systems but excited relatively little further action, largely because the images were poor and the mechanical apparatus was cumbersome, leading to frequent breakdowns in the equipment.

## *Electronic Television*

Sarnoff, keeping a watchful eye on the field, began to feel that the mechanical approach to television was just not going to do the job. At the same time, Sarnoff's reputation in the radio field was growing, and Zworykin decided that he needed the backing of someone like Sarnoff to develop his ideas into a working system. In January 1929, he met with Sarnoff and outlined his ideas.

They felt an immediate connection. Possibly it had to do with the fact that they were both Russian immigrants. They came, however, from very different backgrounds. Sarnoff, a Jewish immigrant who arrived penniless and not speaking a word of English at the age of nine, never even finished high school. (He later took and completed an evening course in electrical engineering at Pratt Institute in New York.) Zworykin came from a well-off family and had a good technical education, including having earned a Ph.D.

More likely, the reason they clicked was that when Zworykin outlined his ideas for an electronic eye, each saw in the other the answer to his specific needs. Sarnoff offered the funds and the business and marketing expertise that Zworykin needed; Zworykin offered the scientific background and technical abilities that would enable RCA to get into this field early and corner the patent market.

Sarnoff quickly worked Zworykin into his plan and made arrangements to set him up at the RCA research center in Camden, New Jersey, along with four engineering assistants. Though the Patent Office had not yet granted Zworykin's patent, mainly because he had not been able to demonstrate that his method could actually work, Sarnoff felt that with his help Zworykin would get the job done.

Still, Sarnoff had heard that Farnsworth was beginning to produce results and was wondering whether the young inventor could spell trouble for him. He suggested that Zworykin visit Farnsworth's laboratory, which was then in San Francisco, and try to see just how far along Farnsworth was.

The small Farnsworth group had heard of Zworykin's work and had high respect for him as a scientist and technologist. Farnsworth, in fact, saw Zworykin as a colleague.

In April 1930, Zworykin spent three full days at the laboratory. One of Farnsworth's assistants even built a sample of their improved camera tube (called an image dissector, for which Farnsworth had filed a patent application in 1927) as Zworykin was watching. Later, Zworykin had several made at Westinghouse and brought them with him to his new Camden laboratory.

Zworykin, confident that he could outrace Farnsworth, reported that RCA should not bother to purchase the Farnsworth operation. If necessary, Farnsworth could always be approached again later. Sarnoff apparently was not convinced and arranged visits by several other RCA people. It was clear that Sarnoff considered Farnsworth a growing problem.

## *Results of the Visit*

The upshot was that no offers were made to Farnsworth, either from Westinghouse or RCA. One reason was that Zworykin's report on the visit and on the camera tube ended up on the desk of Dr. Alfred N. Goldsmith, an official at RCA, who was still convinced that mechanical scanning was the way to go.

Unfortunately, much about the visit remains unclear. We don't know for sure how Zworykin represented himself, but the records seem to indicate that the Farnsworth group believed he was still at Westinghouse. Nor do we know how much technical benefit he gained. We do know that he moved into overdrive at his RCA laboratory and that on May 1, 1930, just a few weeks after his visit to Farnsworth, he applied for a patent on an improved camera tube. Although the timing is suspicious, in truth the design was different from Farnsworth's. Farnsworth was depending on a purely magnetic method to control the camera beam, which was turning out to be problematical. Zworykin used an electromagnetic method, which was a definite improvement.

Another drawback of the Farnsworth design was its low sensitivity; the image his system could build up was rather weak. Zworykin designed a kind of storage capability, which produced a considerably brighter picture.

So there was improvement in Zworykin's system, but he still had a long way to go. The promise of all-electronic television kept Sarnoff investing, and kept the Camden lab in business. In fact, RCA's public relations department began claiming development of an all-electronic system. It was basically true, but the publicity didn't bother to spell out just how poor the physical results really were. Still, having been apprised of what Farnsworth had done, Sarnoff undoubtedly felt his expensive lab could quickly do it, too, and he felt he had to establish his credentials in the all-electronic field early.

He was hoping to set up RCA with the same strength in television as it had in radio. Imagine what was at stake. If he could manage this, he would have control of the entire world of broadcasting.

But the name of Farnsworth kept popping up as a potential problem.

## *Farnsworth*

Born in a log cabin in 1906, Philo T. Farnsworth knew early that he was meant to be an inventor. Precocious and handy, he announced at the age of six that he would follow in the steps of Thomas Edison. What he couldn't know was that the age of the independent inventor was fast waning and that the man with whom he was to cross swords later would have much to do with this change.

By age twelve, Farnsworth was repairing the electrical machinery on the family farm. He constructed a washing machine for his mother, having wound the needed electrical coils himself. His working knowledge came mainly from a stash of technical and semitechnical magazines that had been left by an earlier tenant. Even the ads, many of them from RCA, fascinated him. Another source was the generator repairman who had to be called in now and then and whom Philo would pump for information. Family legend has it that Farnsworth would deliberately disable the device to gain more time with the repairman.

But he was interested in more than tinkering. He also delved into the basic sciences underlying the devices, down to the electrons that made up the intriguing phenomenon called electricity. Talented in music, he earned some extra money playing the violin at weekend dances, which he used to buy more magazines and reference books.

He read about radio and how to build sets, which was relatively easy to do. He also read about the faltering efforts of the early television pioneers. But when he read about the Nipkow disk, something told the brilliant young scientist that this was not the way to go. As long as the equipment depended on a spinning disk, or its equivalent, the amount of detail that could be picked up and transmitted just seemed to him inadequate to the task, no matter how fast the disk could be spun.

In high school, he was always way ahead of his years, and he practically forced himself upon the chemistry teacher, Justin Tolman, who finally let him sit in on some advanced classes. They became friendly, and Tolman spent a lot of extra time with him, helping him assimilate the basic information he needed.

In the summer of 1921, Farnsworth hitched up the farm's two horses and set to plowing the field for planting potatoes. But his mind was on the challenging question of how to overcome the limitations of the mechanical disk. As he came to the end of a row, he looked hard at it and at the ones he had done earlier, as well. Suddenly it hit him: why couldn't the same idea be used to "paint" a picture on a television tube, row by row.

In contrast to a spinning mechanical device, he knew that a beam of electrons could be magnetically directed, and being essentially weightless, the beam could be moved at incredible speeds. He came up with a conceptual system. The image cast by a lens would create an optical image on a screen. This could be scanned electronically, with each picture element converted sequentially into an electric impulse that is transmitted in order. The image then could be built up at the receiver—element by element, and line by line, dozens of times a second, easily fast enough to paint the needed picture.

As he worked out the ideas, he had to tell someone. He told his father, Lewis, who, though understanding little of what he was hearing, was supportive and advised his son not to spread the idea around.

According to the Farnsworth biographer Evan I. Schwartz, this admonition had a curious undertone. Lewis was not worried about the secret getting out, or loss of patent protection; rather, he was concerned that the neighbors, who already thought Philo a bit odd, would now have proof of his strangeness.[6]

Actually, although just a teenager, Farnsworth was already learning something about the patent system. Reading of a contest in the magazine *Science and Invention* that challenged readers to submit ideas that would enhance the safety and/or comfort of the automobile, he came up with a good idea for a thief-proof ignition switch. With it, he won first prize—twenty-five dollars. But, he learned to his dismay, because the newspaper printed the details, this constituted "prior disclosure" and eliminated the possibility of filing for a patent.

Still, Philo couldn't contain himself and felt he had to talk to someone who might understand what he had done. One day in 1922, Farnsworth walked into the school study hall late one afternoon and looked at the large blackboard at the front of the room. He began filling it with diagrams and equations. Tolman walked in and, noting that Farnsworth was not finished, sat down and waited.

Finally, Tolman was able to ask, "What's this?" Farnsworth laid out the whole thing to the one person he knew who might be able to comprehend what he was saying. But Tolman taught chemistry, not electrical engineering; so to help him understand what he was presenting, Farnsworth drew a simplified diagram on a page from his notebook, tore it out, and gave it to Tolman.

The sketch showed a cylindrical device with a lens at one end that cast an optical image onto a plate covered with a photoelectric compound. Here the image was converted to an electronic image. In Farnsworth's design, the entire image was moved past an electrode that picked up the current from a single element at a time. It was not a complete system, but it was a critical step along the road to an all-electronic television camera and thence to a complete system.

Unburdened by a formal technical education, Philo started building his television career early. With his brother-in-law, Cliff Gardner, he started a small shop, selling and fixing radios. In his spare time, he began trying to create in actuality the television concept that he had outlined to Tolman. But, like Morse and Ford and all other inventors of the time, he had to make everything, including many of his tools, from scratch, or adapt it from other equipment. This included the electron tubes, and Gardner took on the job of glassblower, learning everything as he went.

By 1926, Farnsworth had a small lab in San Francisco and was able to create the beginnings of such a device. He filed his first patent in 1927.

Patent number 1,773,980, covering his electronic television system, was granted in August 1930. A number of articles, both here and abroad, appeared about him and his work, including one in the *New York Times* (December 14, 1930) that described his invention. He also published an article in the March 1931 issue of a new magazine, *Television News,* titled "Scanning Images with an Electronic Pencil." Word was getting around.

## *Competition*

Impatient with the Camden laboratory's progress, Sarnoff decided to visit Farnsworth's laboratory himself, which he did in May 1931. Unfortunately, Farnsworth himself was not there, having been dragged away by court order to testify in a petty lawsuit brought against him by a potential investor. No one can ever say for sure what would have happened if they had met face to face in the lab, but the likelihood is strong that they would have hit it off, and Farnsworth's future might have been very different.

As a result of the visit, Sarnoff made an offer: $100,000 for Farnsworth's patents and his services as well. Though the offer would translate into well over a million dollars today, Farnsworth was not interested. He had been hoping for either investment in his group, or payment of royalties for use of his patents. He turned Sarnoff down flat.

Sarnoff, on the other hand, clearly could have used Farnsworth's technology, but he was not interested in paying license fees. He also was interested in keeping the lid on television for a while longer, so as to be able to milk his radio licenses some more. This he could have done nicely if he could have bought Farnsworth out. It also would have eliminated Farnsworth as a competitor.

Sarnoff knew that his nemesis had nothing like the power and the legal resources of RCA. He still hoped that Zworykin would find a way around Farnsworth's patents, or barring that, that he could fight the patents and perhaps exhaust Farnsworth's resources, at which point Farnsworth would be forced to capitulate.

A few weeks later, on May 6, 1931, Sarnoff announced that he was stepping up RCA's program to develop electronic television, with Zworykin in charge. He also ramped up RCA's publicity campaign, still maintaining that RCA was making progress in electronic television. Yet, a year later, when Sarnoff arranged a demonstration of the RCA equipment for a group of radio executives and engineers, there was still no camera involved. Rather, they used a 35 millimeter film projector and a Nipkow disk to set up the images.[7]

## *Breakthrough*

It's time here to try to determine how much of what Zworykin came up with derived from his visit to Farnsworth's Green Street laboratory two years earlier, and whether Farnsworth had been gulled into giving away his secrets. This has never become clear. Advocates for both sides have very different answers. Farnsworth's brother Lincoln later claimed that Farnsworth had anwered questions "he shouldn't have answered."[8]

On the other hand, Farnworth's backers—more interested in the bottom line than Farnsworth was—had already been in contact with the RCA people, testing out the possibility of some sort of combination. In addition, RCA and Westinghouse were still contractually related (final divestiture did not take place until 1932), so Farnsworth should at least have suspected that the two companies might share information, even if Zworykin was still working for Westinghouse.

The most likely reason for Farnsworth's openness, however, is that by the time of the meeting, Farnsworth had filed over a dozen patent applications covering various aspects of his work, so he likely felt well protected and was happy to show off a little bit.

Farnsworth supporters, pointing to the fact that Zworykin had seen an image dissector tube being built, argue that this shows that Zworykin stole the ideas from Farnsworth. Zworykin claimed at the time that he just wanted to further check out the design. After all, the possibility of financial support, or licensing, was surely in the air, and, said Zworykin, how could they consider this without checking out the tube carefully?

Whatever the true answer, by 1933 Zworykin had made several solid breakthroughs. In 1931, he had come up with what he called the Iconoscope, a camera tube of quite a different design from the one proposed in his original patent application. This is the device that finally led Sarnoff to feel he was on course for cornering the television patent market. Television historian Albert Abramson, in fact, argues that "the disclosure of the Iconoscope marks the beginning of the age of electric television."[9] Zworykin's team also had made a major improvement in a receiving device, which he named the Kinescope. Abramson says that it "changed the history of television for all time. For Dr. Zworykin had produced a simple but ingenious picture tube which made it possible to have a practical receiver in the home of the viewer, a device which the average person could operate, that required absolutely no technical knowledge to run, and could be viewed under almost normal lighting conditions."[10] Nevertheless, the system still needed plenty of work before it could be considered commercially viable.

## *Playing Hardball*

After Sarnoff's visit to Farnsworth's laboratory, Farnsworth managed to work out a deal with one of RCA's few significant rivals, the Philco

Corporation in Philadelphia. Philco set up the Farnsworth group in their own laboratory. Although the facilities were less than impressive, it was a place to work, it provided a solid financial base for their experiments, and the men were pulling in a salary, which had not always been the case.

Though the independent spirits of the Farnsworth group came into conflict with the corporate mind and dress code of the Philco organization, they managed to accomplish their goals, one of which was to establish an experimental television transmitting station for Philco. Things were looking promising, but according to Farnsworth's wife, Elma, Sarnoff began playing hardball. She writes: "An ultimatum was delivered to Philco; either it dumped the Farnsworth Company forthwith, or its license to use RCA's radio patents would not be renewed."[11] After Farnsworth had set up the television station for the company, it did not renew their contract.

In trying to find other alliances, Farnsworth was again stymied. Paul Schatzkin states, "[T]he companies that Farnsworth was approaching were the very companies that paid RCA . . . royalties. And through their own sources, Farnsworth's people learned that RCA had issued another unwritten edict to their licensees: work with Farnsworth, and their radio patent licenses would be terminated."[12]

Donald G. Godfrey, an important Farnsworth biographer, warns against relying too heavily on what he calls family sources. These would include Elma (nicknamed Pem) Farnsworth, and also Schatzkin, who he says "does the best job of presenting the family point of view, [but] is too reliant on the family oral history interviews."[13]

Godfrey argues that the initiative for leaving was Farnsworth's and that the real reason for the separation was philosophical differences between Philco and Farnsworth. Philco wanted to make and sell television sets; Farnsworth was more interested in setting up a patent structure that would include payments from manufacturers for the use of his patents.[14]

Backing up this view is the fact that, to replace Farnsworth, Philco brought in the head of RCA engineering, Albert F. Murray, who was looking to produce a design that was compatible with RCA's patents. He brought with him half a dozen key people from the RCA laboratory at Camden.

Whatever the reasons for the separation, the Farnsworth group was out of Philco by mid-1933. They found new funding and established a new company, Farnsworth Television Incorporated, with offices in San Francisco and Philadelphia. Things once again began to look hopeful, but the Farnsworth group was already in the midst of a major patent fight.

## *Patent Battle*

By 1932, RCA had severed its connection with General Electric and Westinghouse and was now an independent entity. The company, and Sarnoff, therefore had a free hand in dealing with Farnsworth. That same year, they not only issued the ultimatum to Philco, but on May 28, also instituted a patent suit against Farnsworth. The basic claim was that Farnsworth's patent on a "Television System," filed in 1927 and issued three years later, was actually an infringement on the one filed earlier (1923) by Zworykin. Similar to both the Selden/Ford and Wright/Curtiss battles, Sarnoff's attorneys argued that the Farnsworth claims were too broad.

RCA's case was strengthened by the result of an earlier interference suit that had been brought in November 1927 against the Zworykin patent, in which Farnsworth was one of five plaintiffs. That case, concluded in February 1930, went against Farnsworth's side, even after an appeal. The ruling was they they had not proved that Zworykin's system was inoperative, as they claimed.

This was an extremely touchy time for Farnsworth. RCA continued to tout Zworykin as the inventor of electronic television; Farnsworth knew he had to challenge the RCA claim.

He therefore filed a countersuit, alleging that Zworykin's Iconoscope actually infringed on *his* patent. The resulting battle was complex and expensive, and went on for years. For Farnsworth, the war with RCA depended heavily on this patent battle. For RCA, the litigation was a small part of its daily existence.

One of the RCA claims was that no boy of fifteen could have put together the technically complex system required for electronic television. Farnsworth's attorney, Don Lippincott, asked Farnsworth about the early days. Who could back him up? Farnsworth recalled showing his system to his high school science teacher. Happily, they were able to track Tolman down in Salt Lake City. Not only could he recall the instance, but he was able to produce the sketch Farnsworth had given him at the time.

Zworykin, on the other hand, was challenged to produce a physical embodiment of his 1923 claim, one that would perform as his patent claimed; and this he could not, or would not, do. Surprisingly, he turned out to be a poor defender of the RCA claim and wilted under examination by the Farnsworth attorney. One problem was that the Iconoscope as it existed then was quite different, in both construction and operation, from the transmitting device described in his original (1923) patent application; Zworykin undoubtedly felt that showing it at the patent trial would not accomplish anything.

But Zworykin may also have faced another, perhaps even more basic, problem. Elma Farnsworth's description of the background to Zworykin's Iconoscope is worth quoting in full, and *may* have something to do with his inability or refusal to produce a system for the court.

"While the Iconoscope," she wrote, "is most frequently attributed to Dr. Zworykin, its actual origins have never been clearly established. The storage principle [an important part of its claim of improvement on the art] has been traced to the work of a Hungarian scientist, Dr. Kalman Tihanyi, who filed a patent on a similar storage-type tube in 1926. . . . In Europe, Dr. Tihanyi is given credit for this invention. In fact, four of the original claims in Zworykin's Iconoscope patent were taken out and given to Tihanyi. . . .

"So often, we read that Zworykin invented the Iconoscope in 1923, when nothing could be further from the truth. . . . The Iconoscope appears in patent number 2,021,907, filed in November 1931 and granted in November 1935—more than five years *after* Farnsworth's patents were granted."[15]

Is there a bit of family bias here? Perhaps. Albert Abramson, an expert on the technical aspects of the feud, states categorically: "Tihanyi's patents were certainly ingenious and deserving of great credit. . . . However, there is no evidence that he ever built or operated a tube of this kind, and we must give Dr. Zworykin and his research team at Camden the credit they so richly deserve."[16]

In the meantime, although stretched to the limit financially, Farnsworth did manage to put on a very successful demonstration at the Franklin Institute in Philadelphia.

The hearing examiner ruled in favor of Farnsworth on July 22, 1935. RCA, by no means ready to give up, appealed, basing one of its claims on their victory in an earlier case. Farnsworth countered with Zworykin's failure to bring forth a system to back up his claim. RCA appealed again.

The case was again settled in Farnsworth's favor on March 6, 1936. It was a significant victory. Yet Sarnoff would continue to tie up Farnsworth's patents for another three years with more appeals. RCA finally agreed, in 1939, to a settlement of $1 million paid over a period of ten years, for the right to use Farnsworth's patents. It was the only time that RCA, in all of its dealings with firms large and small, had paid for licensing rights.

The importance of the agreement goes well beyond the fact that David beat Goliath, at least in that round. For it brought together the patents of the two leading forces in the field, and thereby paved the way for television as it finally evolved in the United States.

## *Win the Battle, Lose the War*

By 1938, based on their successes in court, Farnsworth felt it was time to gear up for production of his newly protected invention. In December of that year, he and a group of investors formed Farnsworth Television and Radio Corporation. They purchased two major plants that had become available in Indiana and proceeded to equip the plants for a geared-up production schedule.

But now some earlier planning by the Sarnoff group began to pay off in a major way. For while Farnsworth was putting together the new company, Sarnoff and RCA were gearing up to make a big splash at the forthcoming 1939 World's Fair in New York. This was a big one. Described as the World of Tomorrow, it was to be a showcase for new technologies being touted by all the major companies. RCA, like General Electric and General Motors, had built a large, freestanding exhibition hall.

Ten days before the fair opened, in April 1939, Sarnoff appeared at the dedication ceremonies for RCA's expansive, and expensive, exhibition hall and sent off an important salvo to convince the world that RCA and Sarnoff were the only names anyone needed to know in the world of television. On opening day, RCA televised the ceremonies that featured President Franklin D. Roosevelt and Albert Einstein.

The patent battle had not been much of a problem for RCA, but it had dredged seven years out of the seventeen-year life of Farnsworth's major patent. It had cost him dearly, not only in funds, but in the time, energy, and resources he could have used to bring his ideas to fruition.

Neither principal was present when the final court decision was handed down. Sarnoff was busy with his many other interests. Farnsworth had retreated to his country house in Maine and was dealing with some health problems.

By this time, World War II was becoming a clear reality. With only a few years of useful life left for Farnsworth's patents, which would be eaten up by the war, he felt that was the end of his television career. Farnsworth Television and Radio built him a laboratory at his country home, where he could continue to work on projects that interested him.

But by now, all of the nation's facilities were focused on defense activities. World War II did arrive. RCA went to work for the government, as did Farnsworth Television and Radio. So did Sarnoff and Farnsworth.

Sarnoff later was put in charge of setting up the all-important communications network for D day—the Allies' invasion of the Nazi-occupied territory. Then there was more work in Europe and Africa. For his efforts, plus earlier time he had spent in the reserves, he was promoted to the rank of brigadier general. From then on, he preferred to be addressed as General Sarnoff. It didn't hurt his image in his continuing efforts at recognition.

When the war ended, RCA went on to become a power in the television field. This was not an easy task. Getting the field to the point where commercial television became a reality—involving design, manufacture, broadcast, and creating a market—was still a major task. The standards—with each manufacturer pushing its own—were a major stumbling block. Sarnoff, in his usual bullheaded fashion, began production before a standard had been decided upon by the government. But once again, his gamble paid off when the system deriving from Zworykin's and Farnsworth's patents formed the basis for the broadcasting standard that finally was decided upon and which is still in use today.

In the meantime, Farnsworth's world had slowly disintegrated; he had health problems and began to drink. No longer able, or willing, to fight for his rights to the credit due him, he continued to fall further behind in the world of television, as did the company he had left behind. Although it did look at first like the company might turn into a paying proposition, it couldn't buck the power of the competition, and especially RCA, and Farnsworth Television and Radio Corporation was eventually purchased by International Telephone and Telegraph Company.

## *Inventing History*

As Farnsworth's patents did, indeed, begin to run out, and as RCA flourished, its public relations department continued its campaign to establish Zworykin as the actual father of television and Sarnoff as "the clear-sighted and steadfast godfather."[17] This served Sarnoff's objectives just as well as being the father himself and was perhaps an easier sell. By 1959, the *Encyclopedia Americana,* in its article on "Radio and Television Broadcasting," was referring to "the invention of scanning by Vladimir K. Zworykin."[18]

Zworykin retired in 1954, but Sarnoff went on to ever greater glory—including credit for the development of color television—in name as well as position. History, with the help of Sarnoff and RCA, began its common practice of forgetting completely those who fall, even temporarily, out of the public's eye.

Sarnoff, for example, has been credited with being the telegraph operator who picked up the first faint signals that the S.S. *Titanic* had run into an iceberg, alerted other ships in the area, and remained at his post for seventy-two consecutive hours while coordinating rescue efforts when the *Titanic* went down in 1912. His post was on the roof of Wanamaker's Department Store in midtown Manhattan, and all other telegraph stations were closed down by the government to keep the airwaves clear for the desperate messages going back and forth.

Breathtaking. Actually, the story apparently was put together years later by the RCA spin team. Though based on some facts, it was a considerable exaggeration. Sarnoff was employed as a telegrapher by the American Marconi Company, and he did work on the top floor of the department store. But he was not on duty when the *Titanic* went down, and though he did hurry to his office when he heard of the disaster, his post was one of the ones that were shut down to keep the airways clear. He may have stayed at his post, listening for names of survivors, for that long a period. At the time of the disaster, however, his was not the central station, and none of the major newspapers reporting the story even mentioned his name. As the years went by, however, the enhanced story was included in such books as the Lyons biography[19] and even seeped into such respected publications as Eric Barnouw's history of broadcasting.[20]

Another exaggeration had to do with a pioneering radio broadcast of the highly publicized boxing match between Jack Dempsey, the American boxing heavyweight champion, and Georges Carpentier, the French challenger. The July 2, 1921, broadcast was a great success, and Sarnoff was given credit for arranging it.

But, again, it just didn't happen this way. Schwartz maintains that the bout was arranged by someone called Julius Hopp and was initially reported as such in, for example, the August 1921 issue of *Wireless Age*. Several years later, the story began to metamorphize with the help of the RCA press office, and eventually it came out as Sarnoff's creation.[21]

Early on, the publicity people may have felt it necessary to put a strong spin on the Sarnoff story. As the patent policeman for RCA in the early 1920s, he was gathering notoriety among the press and the public in a decidedly negative vein. It probably seemed like a good idea to build up the Sarnoff name in positive ways as well. RCA used every opportunity to do this, and they were very successful.

Sarnoff's authorized biography, for example, was published in 1966 by Harper & Row, a highly respected publisher. The writer was Eugene Lyons, who was a distinguished journalist, a foreign correspondent, a lecturer, an editor, and a biographer. He also happened to be on the payroll of RCA as a public relations consultant. He signed a contract that gave Sarnoff and RCA full rights to edit the manuscript as they saw fit. This is hardly the route to an objective treatment. Although Lyons is reported to have tried to present a balanced account, any negative information was thrown out, and the book came out as a pure puff piece.

Its coverage of Zworykin spans four pages. Farnsworth is not mentioned at all, and the Farnsworth Company gets a bare mention as one of five RCA competitors.[22]

In another biography of Sarnoff, by Carl Dreher, the RCA magic has taken hold, and once again, Farnsworth is mentioned just one time and again only as one of several early workers in the field.[23]

A decade later, Kenneth Bilby did a major biography of Sarnoff, again published by Harper & Row. He apparently tried to be fair, and there is considerably more coverage of Farnsworth. But the gist of the coverage is seen in his summary: "[Farnsworth] succeeded in creating an enduring niche for himself as the co-inventor, with Zworykin, of the television system that the nation would ultimately embrace."[24]

The National Inventors Hall of Fame inducted Zworykin in 1977, five years before his death. Farnsworth's major contribution was finally recognized, but not until 1984. Farnsworth's Green Street laboratory, too, was finally given its due: in 1981, it became an official historic landmark.

Farnsworth and Sarnoff both died in 1971, Farnsworth in March at age sixty-four and Sarnoff in December at seventy-nine. Farnsworth's *New York Times* obituary, though respectful, was two short columns buried on page thirty-two.[25] In contrast, the death of Sarnoff later that year was page one news in papers around the world.[26] Sarnoff left an estate variously estimated at $1 million[27] to $8 million,[28] while Farnsworth left his family deep in debt.[29]

Had Farnsworth found a way to work with Sarnoff, his life surely would have been a lot easier. Zworykin, who died in 1982 at the ripe old age of ninety-three, was honored, prosperous, and content.

Had Farnsworth lived another fifteen years, he might, however, have gained a bit of satisfaction from seeing RCA finally fall upon hard times and the carving up of Sarnoff's empire. In a wonderful irony, RCA was purchased in 1986 by General Electric—which had been forced by the government to get rid of it half a century earlier. General Electric then sold off the various parts, except for NBC, the broadcasting arm that Sarnoff had created.

Sarnoff's name, however, remained a valuable property. General Electric sold the RCA laboratory to SRI International, and it became a for-profit subsidiary. Now known as the Sarnoff Corporation, it has prospered and is growing into an impressive technology campus in West Windsor, New Jersey.

## *Who Should Get the Credit?*

Does it really matter who gets the credit for being the father of television? It's easiest to say that everyone who worked in the field was important. Or, zeroing in, that it took both Farnsworth and Sarnoff to bring the field to the large-scale, all-important field it has turned out to be.

But, Schatzkin (an ardent Farnsworth supporter) insists, the priority question involves more than pride. He writes, "It matters because the suppression of the true story deprives us of some important knowledge of the human character. It tempts us to believe that progress is the product of institutions, not individuals. It tempts us to place our faith in those institutions, rather than on ourselves."[30]

Note the phrase "suppression of the true story." Godfrey counters, "No one was suppressing the story. The problem is that no one was telling it. The RCA Corporation simply did a better PR job than did the Farnsworth corporations. Until Pem [Elma] Farnsworth started her work the RCA people were the only ones talking. . . . I think today the story is pretty well balanced for anyone who'll bother to look at it objectively."[31]

True, Sarnoff did not invent anything. Like Morse, he took from here and from there; but he did far more. As Bilby puts it: "Although Sarnoff was not an inventor, he was the most successful innovator of his era, with the ability to pinpoint the need for an invention and then flog it through developmental stages to the marketplace."[32]

Further, says the television historian Albert Abramson, "Sarnoff bought up what he needed and never stole any of it. [In] the American system of television . . . that is still in use today . . . very little of what Farnsworth contributed is left." Farnsworth, he adds, "was offered a job with RCA but wanted to run his own ship."[33]

Finally, Sarnoff was always ready to sacrifice immediate profits to achieve a long-range goal. He once pointed out that his firm had invested $50 million (a huge sum in his day) before it saw a dollar of profit from television.

Imagine a major company leader today, in a time when bottom-line thinking rules, staying the course like that.

## *Elma Farnsworth and Friends*

Farnsworth's wife, Elma, who had been a great aid and comfort to him, tried valiantly to keep Farnsworth's name alive after he died. During his long financial and emotional struggle, he had not been interested in writing an autobiography but had dictated his thoughts to Elma. With extensive help from family and friends, she was able to produce a personal biography of her husband; it was put out in 1989 by a small publisher, Pemberly-Kent, in Utah.

In addition, and perhaps as a result of the book, a small cadre of loyal advocates arose. Today, an Internet search brings up a surprising variety of sites, many of which echo the familiar phrase: "We wuz robbed."

His champions appear to be having some success. Perhaps the new century has something to do with it. In 2001 and the first half of 2002, no less than three full-scale biographies of Farnsworth were published. And, perhaps significantly, they were put out by a university press and two major publishers.[34]

What are the chances that sixty years of public relations spin can be unspun? Tune in tomorrow.

# CHAPTER 8

# *Rickover versus Zumwalt (and Just About Everyone Else)*

## Nuclear Submarines and a Nuclear Navy

The military is built on discipline. Those who flout it usually don't last long. Yet Hyman G. Rickover, a small, rude, brilliant, driving, unorthodox Navy man, often treated the U.S. Navy and its rules with absolute contempt. He ignored orders he didn't like, wore his uniform only when and if he felt like it, and acted more like an industrial magnate than a midlevel officer. Yet his Navy career spanned some sixty-three years.

In that time he made many enemies, but had admirers as well. When he was passed over for promotion by the Navy in 1951 and 1952, his supporters went to Congress and had the decision reversed. Facing mandatory retirement in 1964, he managed to beat it off by appealing to presidents every two years until 1982—when he was finally forced to retire by direct order of Secretary of the Navy John F. Lehman, Jr.

But he was a doer. He was one of the early workers on the application of nuclear power, and later, he changed the face of the U.S. Navy forever. He argued that Navy vessels could put nuclear power to work to extend their reach and would thereby have a freedom to roam the seas that went far beyond anything ever seen before. Rickover was not the first to come up with this idea. But through truly superhuman effort, he drove the program that made nuclear-powered ships, especially submarines, an important part of the Navy's fleet.

Squeezing nuclear reactors into submarines—even the larger ones designed and built under his regime—required an amazing amount of technological wizardry. Even the metals used in the reactors and heat exchangers had to be carefully chosen and evaluated. Rickover made things happen that others thought impossible. After much thought and experiment, he and his team decided, for example, that zirconium would be needed for cladding the fuel elements. But zirconium was an exotic metal at the time and had been available only in the tiniest amounts. Now it would be needed in tonnage lots. Reactor-grade stainless steel was another new material that the materials production industry had never been called on to produce. By the time Rickover was finished, a whole new industry had been created, based on standards that went far beyond anything called for in the past.[1]

During World War II, a typical fleet sub could stay submerged for no more than a week, perhaps two with special arrangements. Today, nuclear-powered subs can stay submerged for many months and can even cruise comfortably beneath the Arctic ice cap. Nuclear-powered ships, especially missile-launching subs, play an important part in the Navy's arsenal.

Making this happen required Rickover to step on a lot of toes, in some cases to exert pressure to the point of pain. Further, although many felt that nuclear-powered subs made sense, Rickover tried to use his growing power to promote use of nuclear propulsion on all of the Navy's larger fighting ships.

It was here that the debate escalated into a true tug of wills and expanded to include not only a significant portion of the Navy, but also the Department of Defense, several of the major shipbuilding companies, and a surprising variety of other individuals in both military and civilian agencies.

Rickover certainly didn't get everything he wanted, but today, four out of ten major combatant ships in the U.S. Navy move under nuclear power.

Was he the oracle who turned out, after decades of intense battling, to be right? Or was he a false prophet who, in the long run, did more harm than good?

## *A Malady?*

Although Rickover managed to infuriate an amazing number and mix of individuals in both the military and civilian worlds, everyone recognized him as something special, and he has been described in many ways. One of the more interesting is provided by John Piña Craven, ocean engineer and naval historian, who refers to him in his later years, with more than a little irony, as the Kindly Old Gentleman.[2]

But one of Rickover's fiercest, or at least loudest, opponents was Admiral Elmo R. Zumwalt, Jr. A career Navy man, he was appointed chief of naval operations in 1970 and became, at age forty-nine, the youngest four-star admiral in U.S. history. As CNO, Zumwalt was forced to deal with Rickover, who should, officially, have reported to him. But by clever maneuvering and, perhaps, a lucky set of circumstances, Rickover exerted a power that belied his five-foot-five, 125-pound frame. And so their battles were pretty much evenly balanced. Zumwalt compared their relationship to that of two jungle animals warily circling each other.

## *Rickover the Power Broker*

In the early days of his long career, Rickover had spent time on several ships, including a destroyer, a battleship, and a submarine. In 1937, he finally was given command of the USS *Finch*, a rusty old minesweeper that was now being used to tow naval targets. He quickly earned a reputation as a severe taskmaster.

Later that year, fed up with that command, he applied for and was granted a status known as EDO (Engineering Duty Only). Although it was a coveted and highly competitive assignment, it also meant he was no longer permitted to command a ship, and he was detached as commanding officer of the *Finch*.

On the one hand, this took him out of the usual pecking order of the line officers, which was fine with him. On the other hand, the change made it more difficult for him to make his way up the promotion ladder; but he found a way around this, too.

During World War II, he served as head of the electrical section of BuShips, the Bureau of Ships. BuShips was concerned with the construction, maintenance, and supervision of the Navy's ships and shipyards. (It was dissolved in 1966 and its functions were handed on to the Naval Ship Systems Command, and then in 1974 to the Naval Sea Systems Command.) By 1946, the atomic bomb had shown the incredible power of the atom. In March of that year, the Oak Ridge National Laboratory, which was working in this area, invited the Navy to send a delegation to Oak Ridge to look into the atom's possibilities as a propulsion mechanism. Rickover was among a small group that was sent. He became fascinated by this new energy source and established himself as the Navy's expert on nuclear power. He also worked on some of the early experimental reactors, which led to the first civilian nuclear power reactor, at Shippingport, Pennsylvania.

Later the same year, President Truman set up the U.S. Atomic Energy Commission with the objective of both fostering the application of this new technology, and of regulating of it.[3] Shortly, Rickover's drive and unusual technical abilities, plus some ingenious politicking on his part, led to a unique arrangement, wherein he ended up working for both the Navy and the AEC. The AEC was responsible for the design, development, and safe operation of nuclear reactors.[4] The Navy designed and built the Navy's ships. When nuclear ships were involved, however, the AEC had to take account of the Navy requirements, and vice versa. But running the reactors was a complex operation, so the AEC trained the officers and men who would operate these. The nuclear Navy became a joint operation, with Rickover in charge of both the AEC portion and the Navy portion. Each of these positions would have given him enormous power; the combination was unbeatable.

The AEC was a civilian agency, reporting directly to the President and to Congress through the Joint Committee on Atomic Energy. When he needed the clout of the military command, he had it. But any time he wanted to accomplish something without going through the Navy's chain of command, he simply operated with his AEC hat on.

Clay Blair, Jr., a World War II submariner turned journalist, maintains that, as a result, "Rick was able to write himself letters, to which he would immediately dictate a reply, thus, in moments, obtaining complete Navy-AEC coordination and approval for whatever problem arose."[5] Frank Duncan, a naval historian and the author of *Rickover and the Nuclear Navy,* argues, however, that while Blair's wording is colorful, getting Navy-AEC coordination wasn't quite that easy.[6] Nevertheless, Rickover was in a unique, and very powerful, position.

Rickover's immersion in his work was total: it included interviewing, hiring, and training of personnel—including the crews of the nuclear-propelled ships; setting up an administrative apparatus that was effective and powerful; active involvement in the design and development of nuclear reactors and, by extension, the design and development of the seagoing craft they went into—including both submarines and surface craft; establishing standards for quality control in the manufacturing process at a level beyond anything ever seen before; inventing a scheduling process that permitted building a mock-up of the reactor and its placement in the vessel while staying a step ahead of the actual production process, saving a huge amount of time in the final accounting; and fighting for funding.

Norman Polmar and Thomas B. Allan, who did a major biography of Rickover, say that one of his more hateful practices was a fairly frequent purging of his organization, the Naval Reactors Branch within the research division of BuShips. Did it have to do with his drive for—in fact, his insistence on—perfection? Or was it his way of getting rid of anyone who had the temerity to challenge him? He also would force out people who might replace him.[7]

Duncan agrees that Rickover would get rid of officers whom he considered to be possible replacements for him. On the other hand, some of his senior commissioned officers resigned their commissions so they could continue to serve in the program. As civilians, they were not a threat. Duncan feels, however, that "fairly frequent purging" is an exaggeration, that many of Rickover's personnel stayed with him for decades, and that Naval Reactors was a very stable organization, with most promotion done from inside.[8]

In any case, by 1980, or two years before Rickover retired, he had outlasted six presidents, seven chairmen of the now defunct AEC, fourteen secretaries of Defense, seventeen secretaries of the Navy, and twelve chiefs of Naval Operations.[9]

## *Stodgy Ways*

Right from Rickover's earliest days, he was contemptuous of what he considered the regular Navy's stodgy ways. Most men who came through the Naval Academy at Annapolis, the Navy's main training school, fell in with its regulations and its customs, including the hazing and ridiculing of new arrivals. Rickover never went in for this sort of behavior, and never fitted in. When his classmates were out partying, he stayed back and spent virtually all his free time studying. John P. Craven, an ocean engineer and a naval historian, feels that Rickover "graduated with a lifelong prejudice against his fellow Academy alumni."[10]

By the early 1950s, he already had become a thorn in the side of the U.S. Navy. Blair states, "He would consistently criticize what he called the 'stupidity' of the Navy, the outdated, unnecessary regulations. He would challenge the unwritten rules which made politicking of importance to naval personnel."[11] As Blair put it: "He believed the shortest distance between two points was a straight line—even if it bisected six Admirals."[12]

With his growing power, he pushed the development of nuclear-propelled vessels. As the head of two different organizations, he began throwing his weight around even more strongly. Many people, used to a more typical gentlemanliness in their relationships, were put off by this brash young man.

In July 1951, his name was put up for promotion in the Navy. But the process was by no means automatic. Promotions had to be passed on by a selection board. These nine-Admiral boards, a long-standing tradition in the Navy, operated in secret and kept no minutes of its meetings. It was no great surprise when Rickover was passed over by the board.

But he was getting things done; by August 1950, construction had begun on the prototype of the reactor and surrounding hull for the Navy's first nuclear submarine, the *Nautilus*. The experience gained in its construction was then put to good use when the actual sub itself began to be built; the keel for the full-size *Nautilus* was laid less than two years later, on June 14, 1952. But the prototype paved the way for the sub's unquestioned success and shaved years off what would have required if construction had awaited completion of the mock-up. It was a brilliant solution to a major problem.

Three weeks after the *Nautilus* keel was laid, Secretary of the Navy Dan Kimball presented Rickover with a Navy medal for his part in the Nautilus program.

But the very next day saw an indication of how he was really regarded by many in the Navy. A second nine-man selection board convened to consider the names of captains given to them for promotion to rear admiral. Four EDOs were selected. Rickover was not among them.[13]

This was bad enough. But according to Navy regulations, if a captain is passed over for promotion twice and has completed thirty years of service, he is automatically retired from the Navy. Rickover would have to retire the following June.

Was this a subtle prejudice against technical specialists? Not really; four others had been chosen. There were rumblings that there was some anti-Semitism involved. But it would have had to be aimed at Rickover's Jewish origins, for by then Rickover had married an Episcopalian and considered himself a member of that group.[14]

Rickover's supporters believed strongly, and probably rightly, that if Rickover were forced to retire, his group might very likely be broken up, or would lose its effectiveness. Another more subtle possibility was that young men with talent in science and engineering would steer clear of the Navy if technical people were not properly rewarded.

## *Attack and Counterattack*

Rickover's supporters therefore mounted a strong campaign to force a change in what, to them, was a disastrous decision. By this time, Rickover's unique story had already piqued the curiosity of the press, and this most recent development brought it all to a head. A series of articles began to appear, pointing to what seemed a deep-seated Navy prejudice against this unusual man. Among the crusaders was Clay Blair, Jr. One of Rickover's first and most ardent supporters, he began working on a major article about Rickover for *Time* magazine.

As word of this got around, the Navy, concerned, mounted a counterattack. For example, in an astonishing replay of the Sarnoff/Farnsworth saga, a popular nationwide television show presented Rear Admiral Homer Wallin as "the man responsible for the development of the atomic submarine."[15] There were some underhanded attacks as well, including a rumor that Blair, a Catholic, was part of a Jewish "cabal."[16] Although Blair was concerned about this, he persevered and came up with a strong story, which later grew into his book, *The Atomic Submarine and Admiral Rickover.*

The pro-Rickover publicity campaign, plus some powerful lobbying and behind-the-scenes maneuvering in Congress, had the desired effect. In February 1953, the Senate Armed Services Committee, which had the power to hold up the choices of the selection committee, threatened to hold up the promotions of the thirty-nine captains chosen by the selection board while the committee arranged for an investigation of the entire selection board process.

The Navy, fearing for the very existence of what had become an integral part of the system, immediately surrendered. A special loophole was found in the passed-over ruling which stated that twice-denied captains could be retained on active duty provided they satisfied certain requirements. The Navy set up a specific requirement that one of those being recommended for retention should be experienced in the field of nuclear propulsion for ships. Rickover's name was not mentioned, but it was not hard to figure out who was going to be promoted. He would, in other words, remain on duty until the next selection board was convened, but the board would be bound by this special instruction. In July 1953, Rickover was promoted to rear admiral, and the requirement that he retire was no longer a threat.

But in the entire history of the Navy, no naval officer, or his representatives, had ever publicly challenged the selection board process. The challenge, even though it apparently came from Rickover's supporters rather than from Rickover himself, set back his relationship with the traditional Navy even further.

For example, on August 7, 1958, William R. Anderson, the commanding officer of the *Nautilus,* was awarded the Legion of Merit by President Eisenhower for the ship's wondrous underwater trip from Pearl Harbor to England via the North Pole.[17] This was done at a public ceremony in Washington, D.C. Rickover, astoundingly, was not invited. In answer to a considerable uproar, the Navy insisted it was an oversight. Rickover's wife, Ruth Rickover, referred in a rare public outburst to some of the Navy brass as "stupid windbags" who were out to "hurt my husband for his independent and free spirit."[18] The press picked up on the snub and also began asking how it was that Mrs. Rickover had not been asked to christen any of the first six nuclear subs to go down the ways.

The Navy's dislike of Rickover was certainly not universal. After the ceremony, Commander Anderson was to travel to New York to be honored in a ticker tape parade, but he made it his business to stop off first at Rickover's office to pay his respects.

On the other hand, Rickover's relationship with Congress began to grow stronger and fonder. Senators and representatives were invited for trips in submarines; the first underwater meeting of the Joint Committee on Atomic Energy was held in the *Nautilus* just two months after her launching; the wives of congressmen rather than naval officers were called upon to christen new ships and submarines; Rickover arranged to name nuclear submarines and, eventually, even a supercarrier for members of Congress. For, as Rickover well knew, even if he were retained on active duty as a result of his promotion, he could still be pulled off the nuclear program, and he might very well have to call on Congress to keep him in charge of the program.

As his power grew, he even began arguing strongly against positions stated by such eminences as the secretary of the Navy [a civilian] and the chief of naval operations [a Navy man]. Nevertheless, in October 1958, he got his third star and was promoted to vice admiral. He continued his crusade for a nuclear-powered Navy.

## *Personality Conflict*

Zumwalt became CNO in 1970, but he and Rickover had crossed swords a decade earlier, and their mutual dislike could have begun even before that. Rickover was seventy when Zumwalt was appointed CNO at age forty-five. This, says Frank Duncan, suggests a generation gap.[19] Duncan adds that Zumwalt was no run-of-the-mill academy graduate and already had an impressive career; he was, in fact, the youngest CNO ever appointed.

Patrick Tyler, a well-known investigative reporter who had met and interviewed Rickover many times, points out that Zumwalt had come up through the nonnuclear side of the Navy, and describes him as "a surface ship man who had shunned the nuclear navy because he loathed Rickover's style."[20]

At their first actual meeting, Rickover had interviewed Zumwalt, who was at that time in command of a conventionally powered ship. The job opening was possible command of a ship that was both larger and nuclear powered. By that time, Rickover—who felt impelled to be in on the selection of virtually everyone involved in the nuclear program—had established a procedure that was more of an inquisition than an interview. He left the initial search for technical ability to trusted members of his team. He wanted to see how the applicants would behave under fire. And he often came out swinging, sometimes hard. Zumwalt included a ten-page description of the ordeal in his memoir. Titled *On Watch,* it was published 1976, two years *after* he retired. Here are a few extracts:

> After I sat down, Admiral Rickover looked through the notes that the three [previous] interviewers had made on me. He then looked up and, with a very serious expression which to me resembled a sneer and which expression did not leave him (with two exceptions) throughout the entire period I was with him, said, "Everyone who interviewed you tells me you are extremely conservative and have no initiative or imagination." I sat silent.
>
> Adm. R.: What do you have to say about that?
>
> Cdr. Z: I need a few seconds to reflect on that, Admiral. It is the first time I have received a charge like that about me.
>
> Adm. R.: This is no charge, God damn it. You're not being accused of anything. You are being interviewed and don't you dare start trying to conduct the interview yourself. You are one of those wise Goddamn aides. You've been working for your boss for so long you think you are wearing his stars. . . . You are so accustomed to seeing people come in and grovel at your boss's feet and kiss his tail that you think I'm going to do it to you. Now (turning to Captain Dunford) get him out of here. Let him go and sit until I think he is ready to be interviewed properly. And when you come back in here (turning to me), you better be able to maintain the proper respect. Do you understand?
>
> Cdr. Z.: Yes, sir.
>
> Captain Dunford escorted me to a barren room, apparently referred to as the "tank" . . . [21]

And so on, for nine more pages, all gleaned from notes Zumwalt made right after the experience. From reports made by others who had undergone the process, there is no reason to think that Zumwalt's interpretation was exaggerated in any way.

The upshot? Zumwalt writes: "I left and returned to the office to call my detailer in BuPers [Bureau of Personnel]. I reported that I had completed the interview, that I was sure I had not been selected, that I had set a new record of four trips to the tank, and that he should go ahead and issue my originally slated orders to USS *Dewey* [a guided-missile frigate, later called a cruiser].

"One hour later I received a telephone call from BuPers informing me that Admiral Rickover had picked me for one of the two nuclear jobs."[22] Zumwalt turned down the offer.

Another reason for the underlying animosity: In 1966, Admiral Zumwalt, at the time commander of a cruiser/destroyer flotilla, was called in by Secretary of the Navy Paul Nitze and the then CNO, Admiral David McDonald, to head up a newly formed Division of Systems Analysis in the Navy. Rickover, a shoot-from-the-hip guy, had no patience with either systems analysis or its practitioners. Zumwalt was clearly in this category.

## *Policy Decisions*

But their real battles often involved major policy decisions. Further, Zumwalt's four years as CNO came at a difficult time—for the Navy, for the country, and for him. In *On Watch,* Zumwalt describes several of the major problems he faced when he began this work. The Soviets, in their zeal to become a world power, were putting huge efforts into building up their Navy, while the United States was still licking its wounds from the Vietnam War, and the country was not happy about spending large sums on military buildup.

He also faced an interservice rivalry for the limited funds. "I am not the person to evaluate the extent of my own bias," he wrote, "but I think it fair to point out that following three air CNOs in a row, as I did, I was bound to have some redressing to do."

Then he stated flat out: "A final malady that afflicted—and continues to afflict—the whole Navy . . . can be described in one word: Rickover."[23]

Both Rickover and Zumwalt agreed that the U.S. Navy needed some serious beefing up. But they had diametrically opposed ideas on how to go about it. Though Zumwalt's memoir ranges widely, his battles with Rickover form an important theme, one that erupts here and there throughout the book. By 1970, Rickover was well ensconced in his position(s) and Zumwalt knew he was in for a battle. He wrote: "I knew as soon as I was designated CNO that developing a productive working relationship with Rickover was among the toughest nuts I had been called upon to crack. In my exuberance over being chosen to head the Navy, I believed I could do it. I was wrong."[24]

Part of the problem had to do with Rickover's methods. One of their battles, for example, involved the introduction of a new nuclear-powered attack submarine, the SSN-688. Zumwalt was not crazy about it; he thought it was unstable in tight turns and was just too expensive for what he considered the marginal advantages it did offer. But he knew that the sub would be built, so the question was not whether to build, but how many. Zumwalt wanted fewer than Rickover, so as to have some funds left for a set of "Low" technology ships he was pushing. "Low" was Zumwalt's expression for moderate-cost, moderate-performance ships and systems that could be turned out in relatively large numbers. He felt these large numbers were needed so that the Navy could be in enough places at the same time to get its job done.

"As for the SSN-688," Zumwalt wrote, "like everything in which Rickover has a hand, it had complications leading to ramifications resulting in shenanigans."[25]

In one of Zumwalt's conversations with Rickover, recorded by Zumwalt, Rickover says, "Well, we've been through that and my feeling is that you want to reduce submarines—but that's all right, you can have that feeling—you're CNO—you're boss." But then, Zumwalt complained, Rickover went on to push his own cause in his usual behind-the-scenes manner.[26] In any other case, the boss/underling hierarchy would be clear. Not here: Rickover did everything possible to change the meaning of the word *boss*.

Fortunately for Rickover, our country's government is set up in a way that balances the power of the main branches of government—the executive, legislative, and judicial divisions. In addition, there are also checks and balances between the military and civilian roles. In the United States, the military does not reign supreme as it does in some countries, and it must answer to our civilian leaders. Over the years, Rickover juggled his two hats judiciously, often, and to good effect.

But he also knew that although he could thumb his nose at his own Navy, and in some cases even the Department of Defense, he still needed allies. These he tended to find in the Congress and some of its committees. Both Zumwalt and Rickover lobbied appropriate members of the Armed Services and Appropriations Committees of both houses. Zumwalt complained, however, "House Appropriations, under George Mahon of Texas, tended to listen harder to Admiral Rickover than I thought it should . . . to the considerable detriment of my 'Low' programs for relatively inexpensive ships."[27]

The arguments were strong on both sides. Part of the conflict had to do with a cost/benefit ratio. Everyone agreed that nuclear propulsion offered certain advantages over the standard diesel engine. But nuclear ships are considerably more expensive both to build and to run; they also require more highly trained personnel. Do they offer enough advantages to justify their cost? Zumwalt argued that effective control of the seas requires a large number of ships, which would, realistically, include a large number of "Low" ships. Rickover didn't like that idea at all, and when Zumwalt assumed his position as CNO, several "High" ships were already in the works. These included the SSN-688, the LHA, an amphibious ship; DD-963, an upgraded destroyer/escort; CVANs, nuclear-powered aircraft carriers; and DLGN's, nuclear-powered guided missile cruisers, used to escort the nuclear carriers. The Navy had already signed contracts for them. "The trouble with them," Zumwalt wrote, "was that they were too good in the sense that the Navy had given up too much to get them. . . . [The problem was compounded by an] absence from even a drawing board of any Low types."[28]

Another example of a run-in took place at the end of 1973. "Rickover," Zumwalt wrote, "made a crafty and determined effort to get $244 million of construction money allocated to two of his guided missile frigates (DLGNs). Since that would have meant not building at least eight and probably ten patrol frigates that the Navy badly needed, I fought him tooth and nail. Rickover's repertoire of wiles ranged all the way from the almost sublime one of trying to slip the delivery date of the desperately needed third nuclear carrier by more than a year so that Newport News [shipyard] would be able to make room for DLGN work, to the totally ridiculous one of getting himself invited to the Schlesingers for dinner so he could make his pitch to the Secretary [of Defense] in the privacy of his home on the day after Christmas."[29]

Result: as often happens, a compromise, or so it seemed. After further discussion with Zumwalt, "Secretary Schlesinger informed Rickover that he would add the 100 million onto the budget, provided Rickover promised full active support of the Sea Control Ship and Patrol Frigates [both Low-technology ships advocated by Zumwalt]." Rickover agreed. Then, Zumwalt continued, "Although Rickover got, with my support, his $100 million add-on, which was less than half of what he wanted, and it did not come out of the hide of non-nuclear ships, he did not keep his bargain with Secretary Schlesinger, but continued to work against any and all Low contruction programs."[30]

## *The Trident System*

Even when they agreed, they disagreed. To counter the Soviets' growing and improving submarine fleet, to deal with the growing susceptibility of our land-based missiles, and to give the United States greater leverage at ongoing talks with the U.S.S.R. regarding strategic arms reductions, a new American entry was being looked at—ULMS (underwater long-range missile system).

The craft being designed was given the name Trident. A huge craft, on the order of six hundred feet long that could carry twenty-four updated, longer-range nuclear warheads, it was to replace the aging Polaris system. Polaris dated back to 1959, and was last updated in 1967. The Russians had in the meantime been updating their own fleet and had developed a new line of fast, quiet attack subs that put our existing ballistic missile subs in greater danger.

But once again the basic inclinations of the two men led to clashes. Rickover's official status supposedly gave him dominion only over the reactors going into the naval vessels. In practice, however, he was able to broaden his grasp by pointing out that the safety of the reactors depended heavily on the design of the rest of the ship.

Zumwalt complained: "Rick, as usual, was prepared to capitalize on the opportunity. He had a design for a new reactor ready to lay on the table, a reactor so big and heavy that only a huge hull could accomodate it, and that huge hull, in turn, would have to be fitted with an unprecedentedly large number of missiles to justify its size and cost. Designing a ship from the power plant out never has struck me as an ideal procedure, but practically all nuclear-powered ships have been designed that way because Rickover, the power-plant man, has always been first on the scene with a design."[31]

Rickover never bothered with a point-to-point rebuttal of Zumwalt's many charges, but his name and views appear often in the *Congressional Record.* [32]

Further, there have been plenty of others who have chimed in. For example, Frank Duncan, a well-known and respected naval historian, gives a rather different view of the Trident story:

> Rickover played a major role in the Trident effort, a fact made possible by the success of the technology developed in the naval nuclear propulsion program. The propulsion plant he proposed would have the latest techniques in silencing, allow for the power required for equipment that might be installed years later, and (as Zumwalt later acknowledged) add safety to the operation of the ship. For Trident, Rickover followed his usual practice of constructing a full-scale mockup of the reactor compartment and engine room. Not only did it allow assurance that the shipyard workers would be able to carry out their specialties in the close quarters, but it permitted designers to pretest a new maintenance concept. In this approach large hatches allowed the [later] replacement of questionable equipment.

> [He adds, in counterpoint to Zumwalt's assertion]: It was an error to assert that the propulsion plant set the size of the ship: the dimensions and number of missiles determined that. Rickover had nothing to do with the technical aspects of the missile.[33]

In any case, it was, again, Rickover's design that won out, though there was still plenty of soul- and pocket-searching to be gone through regarding how many were to be made and how fast to build them.[34] The subs, coming in at a cost of about a billion each, fully equipped, began entering the fleet in 1981.

Zumwalt also objected that Rickover's single-minded obsession with nuclear-powered craft blinded him to potentially useful developments in other areas, and that his opposition prevented such developments from proceeding at an orderly pace. In 1967, for example, the Navy was seriously considering the possibility of adopting aircraft-type gas turbines for warships. These power plants could be powered up much faster than nuclear reactors; they also promised easier maintenance and would require fewer and, likely, less highly trained personnel to operate them. Rickover effectively squelched these plans. Not until the mid-1970s were they resurrected and put to work—largely through the efforts of CNO Admiral Zumwalt.

## *Zumwalt Gets Rid of Mickey Mouse*

In one sense, Zumwalt and Rickover should have been able to get along. The dust jacket for Zumwalt's book says, for example, "In four short years, he transformed the Navy, [including] eliminating the Navy's demeaning 'Mickey Mouse' regulations of dress, behavior, and family life."

Spelling this out further, Zumwalt wrote, "Eliminating Mickey Mouse, relieving some of the pressure naval service puts on family life, and creating new opportunities for fun and zest were deeds that were difficult to perform only in a bureaucratic sense. They required much prodding and tweaking of long-established regulations, routines, and mind-sets."[35]

Rickover surely could identify with the prodding and tweaking aspects. But the terms *fun* and *zest* were not part of his vocabulary. In what he expected from his people, he was dead serious, and very demanding.

Zumwalt added, "It amuses me a little that I am known mostly as the CNO who allowed sailors to grow beards, wear mod clothes, and drive motorcycles."[36] It amuses him because, while he did feel that relaxation was important for the morale and well-being of the Navy people, he also believed that he had made far more significant contributions than that to the well-being of the Navy as an institution, and especially, to our military's response to the growing maritime efforts of the Soviet Union.

After Zumwalt's tour of duty as CNO, he chose to retire at the relatively young age of fifty-three and went on to a career in industry, where he became the director of several major corporations.

## *Rickover's Later Years*

Rickover's later years were not nearly as impressive as his earlier ones. In addition to the fact that his obsession with nuclear propulsion probably did prevent further progress in other areas, there were some serious cost overruns at the companies he worked with so brilliantly in the early days. Though he fought them, and although some were fraudulent, overruns did happen on his watch. He was also called up on charges of receiving expensive gifts from certain contractors. In usual fashion, he never denied the charges, but maintained that there was nothing wrong in what he had done.

Toward the end, Rickover himself began to have second thoughts about nuclear power. His main fear was of nuclear weapons, but by extension, he began to question his quest for a nuclear Navy as well. The nuclear Navy is still alive and well, but as the nuclear ships age and come up for decommissioning, the same arguments about costs and benefits flare all over again. What will happen in the long run without the powerful push of Rickover remains to be seen.

After his death in 1986, the *New York Times* wrote, in addition to a full obituary, an editorial about him. After listing some of his accomplishments, the editorial stated, "But he stayed active too long. . . . He demanded the Navy's best people, but tried to dominate, sometimes even humiliate them. . . . His mark on the Navy would have been greater had his career been shorter, yet it remains indelible."[37] It was a sad commentary on an extraordinary man.

## *The Long View*

During Rickover's long service, he garnered many accolades, many medals, many decorations. And as the *Times* quote suggests, even among his detractors there is always grudging admiration.

Among all of Rickover's accomplishments, the submarine program is surely his main legacy. For it and the whole nuclear Navy have amassed an amazingly successful record of safety. Duncan's summary of Rickover's program is worth quoting:

> His key to getting strong support was the excellence he demanded and built into Naval Reactors, the research and development carried out at Bettis and Knolls [Atomic Power Laboratories operated by Westinghouse and General Electric, respectively], the procurement activities of the Plant Apparatus Division and the Machinery Apparatus Operations, and the countless contractors and vendors. Weighing heavily in congressional opinion were the reliability of the propulsion plants and the superbly trained officers and men. When so many military programs were in trouble and the results uncertain, the naval nuclear propulsion program was a welcome exception.[38]

Even the objections about Rickover's standing in the way of further progress may need some leavening. One of the current pushes in the Navy today is for all-electric drive, an integrated electric power system, which is simpler, smaller, and more reliable than the conventional combination of electric and mechanical techniques. Yet Rickover fought for an electric drive system way back in the 1960s; it was for a sub called the *Glenard P. Lipscomb,* which was commissioned in 1969.[39]

As for Rickover the man, Edward L. Beach, a retired Navy captain, states in the foreword to Duncan's book, "The U.S. Navy cannot stand many men like Rickover in a single generation, but once in a great while, in a situation of transcendent importance, such a person is needed. Even Rickover's faults, great as they actually were . . . somehow contributed to the extraordinary success of what he accomplished."[40]

And in his memoir, even Zumwalt had to admit: "I have taken occasion to dwell on the ways I think Rickover's long tenure has affected the Navy, but one thing no one can say about him is that he ever sponsored a lemon. Without exception his products have been excellent—too excellent, [as] I believe I have mentioned several dozen times."[41]

CHAPTER 9

# *Venter versus Collins*

## Decoding the Human Genome

The medicines available today would have seemed like miracles a short century ago. Yet virtually every one of them also has side effects that have proved disastrous to those unlucky enough to be allergic or susceptible to them. The dose too must often be adjusted by trial and error. Imagine being able to take a single blood sample and from it determining whether the patient has an aberrant gene that could cause a negative reaction to it.

Imagine being able to test for genetic predisposition to the major causes of illness and death, and thereby being able to produce new therapies for them.

Imagine . . . The list goes on and on.

But until recently it was always "imagine." This kind of medicine has long been nothing more than wishful thinking because the basic information—the set of instructions that determine just how one's cells would respond to biologically active substances—was a great unknown.

Then, on October 1, 1990, amid considerable fanfare, the National Center for Human Genome Research (NCHGR) announced the start of a major scientific program.[1] Designated the Human Genome Project (HGP), it had as its ultimate objective to decode the complete set of instructions hidden in the 3 billion letters that spell out the "Book of Life" contained in our DNA. Projected to run fifteen years at an estimated cost of $3 billion, it was to be a massive, international effort, with the eventual participation of scientists in England, Germany, France, Japan, and China.

The technical term for what it hoped to achieve is the *sequencing* of the human genome. But this was a distant goal. Initially the genome would be mapped, meaning that the locations of the genes among and along our chromosomes would be established, along with a set of markers along the genes. This would make it easier to place accurately the vast number of sequences, or gene fragments, that would be found in the later stage.

Chosen to lead the HGP was none other than James Watson, the codiscoverer of the structure of DNA, which gives some idea of the importance attached to the program. His worldwide reputation and outspokenness would, it was expected, go far to help the program along.

But trouble was already brewing. In April 1991, Bernadine Healy was brought in to head the National Institutes of Health (NIH) of which the NCHGR was a part, which meant that Watson now reported to her. By early 1992, Watson and Healy were already at odds over several issues. One of these had bedeviled the world of genomic research right along, and continues to do so today. Healy strongly supported an NIH decision to apply for patents on hundreds of gene fragments (portions of the code) that had been produced by a then little-known NIH researcher named J. Craig Venter.

Eventually, Venter hoped, these fragments could be put together into complete genes. In the meantime, they were just fragments and gave little understanding of the gene's operation. If nothing else, the NIH reasoned, the applications would help clarify whether the Patent Office would grant patents on genes, or even parts of genes, with no known function.

Part of the uncertainty about their patent status is that genes identified in this work are not in their natural form. They've been copied, spliced into bacteria, or otherwise changed. So in the eyes of the patent examiner, a gene is really a kind of laboratory construct. These constructs can, hopefully, be adapted later for medical use. At that point, there is much less questioning of their patentability. Since the 1970s, when the Patent Office began issuing patents on human genes, these patents have become the foundation of the explosive biotechnology industry. Examples of products produced in this way include insulin, human growth hormone, and erithropoietin,[2] with billions of dollars in combined sales.[3]

The fear among other researchers was that if the Patent Office did grant patents on genes whose function was either unknown or uncertain, the applicants might be able to lay claim to large portions of the genome, just hoping that at some point someone would figure out their function, at which point they would have the territory locked up.

Using one of the first commercially available DNA sequencing machines, Venter was able to sequence data on hundreds of genes simultaneously, whereas standard techniques could deal with only one gene at a time.

The machine that made this possible had been developed half a dozen years earlier, but was still finding its way in the biotech world. Venter, looking for some way to speed up the gene discovery process, met with Michael Hunkapiller of Applied Biosystems, a new firm set up to develop and market the automated machine. (Hunkapiller and the machine's other inventor, Leroy Hood of the California Institute of Technology, had tried to sell the idea to nineteen different companies without success.) Although its method of automated sequencing was clearly faster than the previous manual methods of DNA sequencing, what it produced was vast numbers of gene parts that had to be put together later into some semblance of order.

In 1986, Venter agreed to test a prototype of the machine at his lab. It was a momentous decision. By 1991, he had, using the machine, developed a new approach to gene finding. The problem had long been that the human genome contains long sections of "junk" DNA, regions of the genome that are inactive, but that were being sequenced anyway in the standard methods. Venter was seeking a way to concentrate on the active portions, which constitute only about 3 percent of the entire genome. The essence of the new process involved what he called expressed sequence tags (ESTs), which are bits of expressed genes, meaning genes that are actually doing something and so are obviously active portions of the genome.

## *Automated Partial Sequencing*

Venter was doing his research at the National Institute for Neurological Disorders and Stroke, so was most interested in genes having to do with the brain. On June 21, 1991, he and his group published a stunning article in the journal *Science*. In it, they described a study involving "automated partial DNA sequencing" conducted on more than six hundred randomly selected human brain DNA sequences. The objective was to generate ESTs, which in this case would be gene markers for genes found in the brain. The key operative terms here are *automated* and *partial sequencing*. As the authors stated later in the article: "The EST data, in conjunction with physical mapping, will provide a high resolution map of the location of genes along chromosomes."[4]

Note the phrase "in conjunction with physical mapping." The two sides of the coin were being stamped early. On one side was Venter's route: automated sequencing first and mapping later. On the other was the route espoused by the NIH's HGP: map first and sequence later.

To Venter's credit, his group had spelled out the identity of over 330 genes active in the human brain. At the time, the entire known collection of sequenced genes totaled less than ten times that many.[5] This was an impressive accomplishment.

In their more expansive moments, the EST proponents argued that their approach would be a faster, cheaper way to analyze the human genome and should be pursued instead of the HGP's. This idea got a lot of publicity and was seen as a threat to continued HGP funding. But, said the HGP side, it was a flawed argument, and Watson and other leaders in the field were highly critical of Venter's work. One key aspect of the criticism was that ESTs would not suffice for, first, finding all the coding sequences in the genome and, second, finding all of the functional sequence elements in the genome. In other words, it would miss important aspects of the genome later on.

Watson and company were even less thrilled by Venter's pronouncements, including the statement that he could beat out the entire HGP. As for the patent angle, Watson argued that it was "sheer lunacy" to patent such incomplete information, and added that the automated sequencing machines "could be run by monkeys."[6] Where, in other words, was the inventive step?

Watson was pushed over the edge when Bernadine Healy not only wouldn't back down on the patent issue—the NIH was now trying to patent even more of Venter's ESTs—but invited Venter to play a larger role in planning the research, while at the same time clamping down on Watson's criticisms. As it turned out, the Patent Office turned down the first set of applications in August 1992, but Watson had already made up his mind to leave.

In addition to Venter and the patent problems, there were questions having to do with a portfolio of shares he held in certain biotech companies.[7] His position having become "untenable," Watson quit abruptly in April 1992, leaving the HGP in a precarious position. There were few scientists with Watson's knowledge, stature, and political capabilities. Among the front-runners to replace him was Francis Collins, the head of a well-respected genomics research laboratory at the University of Michigan.

## *Collins*

Trained in both physical chemistry and medicine, Collins had already built a successful career in genetics at the University of Michigan, which had a strong program in this area. With techniques that he had helped develop, he had been instrumental in finding two genes that had tantalized genetics researchers for years, those for the hereditary diseases cystic fibrosis and neurofibromatosis. He also contributed to another major gene discovery, having to do with Huntington's disease, a fierce neurodegenerative disorder. Some hereditary diseases involve several genes that can be spread out over several chromosomes.[8] In the Huntington's case, a mutation in a single gene can trigger its development. Still, finding the gene and the specific mutation proved enormously difficult and took more than a decade. As with so much genetic work performed up to that time, the search went from the disease to the gene. It was necessary to find a variation in the gene sequences being studied that would track with the Huntington's gene, and this meant seeing whether all victims of the disease showed the same misshaped sequence.

Healy invited Collins to lead the new NIH project. He promptly accepted. Why would he leave one of the prime genetics laboratories to head a project with an uncertain future and take a cut in salary as well? This was to be the most ambitious, expensive, and medically significant biological research program ever undertaken, and it would be in his own field. Collins felt, in fact, that "there is only one human genome program. It will happen only once, and this is that moment in history. The chance to stand at the helm of that project and put my own personal stamp on it is more than I could imagine."[9]

## *HGP*

Over the next few years, the program progressed about as planned. By April 1996, the consortium had produced the full sequence for brewer's yeast, involving 12 million base pairs (letters), and roughly 6,000 genes. As hoped, the consortium had also identified many genes that were involved in hereditary diseases, including Alzheimer's disease, muscular dystrophy, and cancer. They were able to do this because they were mapping first and sequencing later. That is, before sequencing the DNA segments, or clumps, of DNA data, the HGP investigators had determined where on the chromosomes they belong.

Like a train lumbering up an incline, the HGP continued along its route—having by 1998 come to the halfway point and having spent $1.9 billion of its projected total $3 billion cost. But because the map-first route had taken up most of the project's time, only 3 percent of the genome had been sequenced. Even though many of the HGP labs were by now also using some of the automated sequencing machines, it appeared that at the current rate, there was little chance of meeting the original schedule. The HGP plans, however, projected significant increases in production rates after the initial pilot phase and, says NIH's Mark Guyer, was always on track to meet the original schedule.[10]

Suddenly, on a parallel track, the Venter express came roaring by.

It came as a rude shock.

## *Venter*

At the same time that Watson was in the process of disengaging from the HGP, Venter, unhappy for other reasons, was also in the process of disengaging. Healy had left with the change of presidential administrations. Without Healy behind it, Venter's application for an expanded role in the project was not rated as highly as other applications in the standard review process and was pigeonholed.[11]

At the same time, he was approached by a medically oriented venture capitalist, Wallace Steinberg, who felt that by the end of the twentieth century, genetic information would be the foundation on which new pharmaceuticals and diagnostic procedures would be built, and he wanted to be in on it. With an initial investment of $70 million (later increased to $85 million), spread out over ten years, he and Venter created two organizations.

The Institute for Genomic Research, or TIGR, would provide space and funds for Venter to carry out his research, and would give him almost complete freedom to publish his findings.

The key word is *almost.* For Steinberg, this was an investment, not a philanthropic contribution. So he set up a separate company, called Human Genome Sciences, which would market the discoveries and developments sure to come from TIGR. Venter initially asked for full freedom to publish the results. They finally agreed that Venter would give HGS six months to review any data that were to go out for publication and twelve months if there was any likelihood that the data could lead to a new drug. Venter accepted. A month later, he promised to release data every three months.

In the initial plan, TIGR would rely on thirty ABI 373A automated sequencers, plus the software and hardware needed to pull the data together. HGS, the business partner, would, they expected, emerge as a special kind of pharmaceutical company in the long run. In the meantime, it began selling some of TIGR's results, including access to Venter's EST data, to several major pharmaceutical firms. In 1993, one of these firms, SmithKline-Beecham, even invested heavily in the new company.

TIGR emerged rapidly as a major player in the evolving genomics field. In 1994, Venter applied to NIH for funding to test his method on the medically important bacterium *Haemophilus influenzae.* NIH eventually turned him down, calling what he proposed "impossible."[12] But Venter, ever in a hurry, had already begun the process on his own, and a few months after the NIH turndown, he published the complete sequence. It was the first completed sequence of any free-living organism, and it started what has turned out to be a microbial sequencing revolution.

In September 1995, the prestigious journal *Nature* published a major paper called "The Genome Directory," along with a 379-page supplement. In it, Venter and over ninety coauthors—from both sides—presented what could be called a preliminary consensus on the human gene collection.[13] On his own Internet database, titled "TIGR Human cDNA Database," he included information gleaned from the public HGP. These data, added to his own major contribution, resulted in the world's largest collection of identified (not fully sequenced) genes, amounting to perhaps half of all human genes.

In 1993, Perkin-Elmer, an old-line manufacturer of scientific instruments, had bought out Applied Biosystems. By 1998, Michael Hunkapiller had had a dozen years to refine his sequencing system. His latest creation, the PRISM 3700, fitted in nicely with the Venter/Steinberg program. So confident was Hunkapiller of its capabilities that it was he who suggested it be used for decoding the human genome, even though the human genome was fifteen hundred times larger than the bacterial genome that Venter had sequenced with Hunkapiller's earlier machine.

At the same time, Venter's relationship with the HGS management was souring. There were quarrels about the required delays in publication and on the patent front as well.

Venter resigned from HGS in 1996 and, along with Steinberg, formed two subsidiaries of Perkin-Elmer in May 1998. One was a new firm, as yet unnamed, which would carry out the sequencing, while Hunkapiller's Applied Biosystems would provide the machines—several hundred of them.

It was all done under a heavy cloak of secrecy. The first Collins knew of it was an invitation to a meeting on May 8 with Venter at the United Airlines Red Carpet Club at Washington's Dulles Airport. Venter suggested that he and Collins share their data, which of course would have meant sharing the credit at the end of the project. This seemed highly presumptuous to Collins, who said he needed time to consider. But Venter was in no mood to dawdle.

Two days later, the story exploded on the front page of the Sunday *New York Times*.[14] Venter announced that his new firm would sequence the entire human genome by 2001, or four years before the HGP target date, and at less than one-tenth the cost. To establish the efficacy of his process, he would start off by sequencing the genome of another important biological research model, the fruit fly *Drosophila melanogaster*. With a genome fifty times larger than the bacterium he had done earlier, but about a thirtieth the size of the human genome, it seemed an ideal bridge between the two. Equally important, it had many genes in common with human genes.

There was an important "but," however. Venter would be using his shortcut method to achieve his final result. This was his "whole-genome shotgun" approach. As we saw earlier, he had tested it in 1995 and had successfully produced a complete genome for the bacterium *H. influenzae*. Again, instead of mapping the genome first and sequencing its parts later, as the HGP tended to do, he would leave for later the time-intensive and more laborious mapping process and would begin with a kind of random sequencing.

But even the new automatic sequencing machines could handle only little pieces of genome, on the order of 500 to 600 DNA letters at a time. How do you construct a sentence that is over 3 billion characters long when you can only handle 600 letters at a time?[15]

Venter's method chemically cuts apart the chromosomes that make up the genome—using multiple copies from several individuals—into a set of randomly cut pieces of somewhat more than 600 letters each.[16] Each set, called a clone, is small enough to be handleable by his new sequencing machines, but long enough to have ends that will overlap with the next appropriate clone in the series. Later, these overlaps would be sought out and matched by pattern-matching software, plus the huge computing complex that was being established at the company's headquarters in Maryland. The matching process is carried out over and over again until the entire genome—containing over 3 billion bases—is in order. The ultimate objective is to fit all the parts together in "sequence," like a necklace of beads.

The final step in the genome sequencing process is often called finishing, meaning filling in the gaps, and correcting the expected errors, ambiguities, and inconsistencies How well this would go later on loomed as a major question in his process—which has not yet been completely answered.

Venter was not the first to use the shotgun method. It actually had been developed several years earlier and offered to NIH, but was turned down as too expensive; scientifically flawed, in terms of its ability to deal with repeated sequences; and too risky.[17] Now, suddenly, the combination of automated sequencing and the shotgun method loomed as a major threat to the Human Genome Project.

## *Criticism*

Venter's blockbuster announcement came just before the annual May meeting of prime genome researchers at the Cold Spring Harbor Laboratory in New York. Most of the attendees at that meeting were sympathetic to the position held by the public consortium, and the criticisms flew—both during Venter's long-planned talk and behind the scenes. One was that Venter would be using the work done at HGP—posted every night and freely available on the HGP's GenBank Web site—and yet would soak up most of the credit if he did indeed come in first. Others feared that he might even end up sticking clones together in the wrong order.

He was called the "Bill Gates of biotech."[18] Kevin Davies, who wrote a solid book on the history of the genome battle, says that Watson "compared Venter's assault on the genome project to Hitler's annexation of Poland."[19]

Watson later denied making such a comparison, but did admit to accusing Gerald M. Rubin, a professor of genetics at the University of California/Berkeley—who had agreed to work with Venter on the fruit fly project—of "collaborating" with Venter.[20] This might have been just an unfortunate choice of words, but it was a word that had been used with distinctly negative connotations during the World War II period.

Maynard Olson, an early leader in genetics, told a House Science Committee a month later that the whole-genome shotgun route was a slapdash effort, full of hustle and PR. He predicted that the method would leave over 100,000 serious gaps rather than the 5,000 that Venter anticipated.[21] Although Venter's method had worked with the microbial genome, the human genome not only is far larger, but contains many "repeat" sequences that may not have the unique identifiers that would be needed for good identification. Olson also predicted that the HGP would now have to lower its standards in order to keep pace with Venter.

Some would say that's just what happened. But, says NIH's Larry Thompson, "It isn't correct to say that HGP had to lower its standards. It did, however, have to change the sequence of its goals."[22] The initial HGP plan was to produce a completed version of the genome by 2005. But at the 1999 Cold Spring Harbor meeting, the leaders changed its initial focus and began working toward producing a "rough," or "working," draft of the genome by spring 2000, which was earlier than Venter's first target date. So what they really did was add an intermediate goal that would provide an important milestone at the same time that Venter had said he would have a milestone.

The HGP would also try for a finished version in 2003. To help meet the earlier deadline, it changed its basic procedure. It would still do mapping first, but would use a combination of human and bacterial chromosomes as a way of speeding up the process.

Three months after the blockbuster announcement, Venter finally came up with a name for his company that satisfied him: Celera Genomics—from the Latin *celeris,* meaning swift or quick.

Things did start moving faster. By October 1999, Celera reported having sequenced its first 1 billion bases. The HGP, having stepped up its production, came in with its first billion just a month later. Mark Guyer insists, however, that these are "apple and orange numbers. What Celera produced was one billion 'raw' bases, i.e., data directly off the sequencers. The public consortium's claim was the production of one billion of *draft* quality, the equivalent of four billion raw bases."[23]

Some say the speedup is attributable to Venter's challenge. But Eric Lander, head of the Whitehead Institute at MIT, one of the mainstays of the public program, maintains that the speedup had all been planned before, that the previous three years had been a "pilot project phase that was devoted to developing the methodology for how to sequence genomes. That phase," he insists, "came to an end in March of 1999, and we went from a pilot operation to a production level in excess of 15 billion nucleotides, or DNA letters, per year. We scaled up 20-fold over the course of about nine months."[24]

## *Scientific Research or Business Opportunity?*

While some of the scientists were worrying about the quality of Venter's results, others were worrying about public, and scientific, access to these results. The world of genomics was becoming increasingly privatized. The Patent Office had already issued more than eighteen hundred patents on complete gene sequences (not portions) that had been laboriously sequenced earlier. Most were for plant genes, but there were increasing applications for human-gene patents, many by pharmaceutical companies. The commercializers argued that this would facilitate the process, using the standard argument of drug companies that the costs of putting a drug on the market is so high that none would get involved if they couldn't put a lock of some sort on the product. The scientists feared that such locks would inhibit their research.

An even more disturbing possibility was that NIH, fearing some sort of political or taxpayer backlash, would even back away from supporting the project altogether. Dr. Michael J. Morgan, the program director for the Wellcome Trust in Great Britain, stated flatly, "To leave this to a private company, which has to make money, seems to me completely and utterly stupid."[25] To back up this position, the trust doubled its contribution to the Sanger Center near Cambridge, England, allowing it to sequence a full third of the genome, up from its originally scheduled one-fifth.

Venter, well aware of these arguments, had promised to release his basic human genome sequence, though in three-month chunks, rather than nightly, as the HGP was doing. Not included—for clear commercial reasons—would be its interpretation of the data, which Celera planned to sell, along with the patent rights to perhaps two hundred to three hundred "novel gene systems," which it planned to patent as the plan proceeded.

There was both grumbling and doubt that Venter would keep his promise to share much of his data. He successfully completed his fruit fly program by March 1999 and promised to release the data to GenBank, thus assuaging some of the doubt and fears still percolating in the public domain. Guyer points out, however, that the data weren't actually released to GenBank until the fruit fly paper was actually published, in March 2000.[26]

Venter, now a major businessman, was well aware that Celera had to make a profit if it was to stay in business. He insisted, however, that he could balance the apparently conflicting objectives of releasing data freely and making a profit. Over the first several years of Celera's operation, he was able to raise a billion dollars in private capital.

## *When Is a Race Not a Race?*

The science magazine *Science News* had spelled out Venter's entry as "modern biology's equivalent of the fabled race between the tortoise and the hare—except that the prizes at the finish line are the priceless secrets of the human body and the tortoise may not repeat its legendary victory."[27]

Throughout the short history of this competition, both Venter and Colins have tried, or have said they were trying, to work in harmony with the other. Right at the beginning, Collins had said he planned to integrate his program with the private company's initiative.[28] That didn't last long. He also referred to Celera's offer to share its data with the scientific world as "disturbingly ambiguous."[29] What he had in mind, of course, was that Venter would have the benefit of all the public research, while the reverse would not be true. Would this matter if this was not a race?

Venter seems also to have been of two minds on the subject. In 1998, he stated, for example, "Francis is not a competitor. He's a government bureaucrat whose job is to hand out money to help get the genome done."[30] Yet by June 2000, Venter (an ardent sailor) had this to say: "They're trying to say it's not a race, right? But if two sailboats are sailing near each other, then by definition it's a race. If one boat wins, then the winner says, 'We smoked them,' and the loser says, 'We weren't racing—we were just cruising.'"[31]

If it was a question of cooperation versus competition, competition won out every time.

There had been earlier attempts at reconciliation, including a meeting on December 29, 1999, but they always seemed to end badly.[32] One point of contention, among many, was, in Collins's words: "Will the sequence of the human genome be freely accessible without restrictions of any sort to researchers in the private and public sectors, or will it not?"[33] Another: After the December meeting, Collins wrote to Venter stating his feeling that it would be unethical for them to publish their data on the human genome without the approval of scientists who contributed to the public database.

Leaders in the genomics field would, in fact, have liked to see a collaboration, but it just never seemed to happen. In truth, it's not hard to understand why. The public group—involving over a thousand of the biotech world's best and brightest—were willing to work as part of a team, but they had the belief that they would go down in history as the team that decoded the human genome. Now they were being blindsided by a latecomer.

On the other hand, right at the beginning of the battle, Venter had suggested that the public group had chosen a "flawed strategy" that would produce a seriously incomplete DNA sequence.[34] Now, Lander feared, naive onlookers—including taxpayers and public funders—might be fooled into thinking that Celera had actually sequenced the entire human genome, when in fact it had produced less data than the public group.[35]

For Venter, being the sort of man he is—fond of breaking the rules and battling doubters—and after taking so many barbs, it would be hard not to be prickly.

Referring to the two leaders, Eric Lander suggested: "At a certain level, it's just boys behaving badly. It happens to be the most important project in science of our time, and it has all the character of a schoolyard brawl."[36]

## *Fix It!*

The significance of the situation entered the political world as well and percolated right up to the top. President Clinton felt that the competitive atmosphere was both counterproductive and unseemly. On April 7, 2000, he sent a note to his science adviser Neal Lane, saying, "Fix it . . . make these guys work together."[37]

A month later, on May 7, Collins and Venter met. It was not an easy meeting. One attendee, Aristides (Ari) Patrinos, later recalled that he had "never seen them as tense as they were that day."[38] Nevertheless, they met a second time and a third. A major factor in the discussion was the fact that Venter was anxious to announce a momentous development, the completion of a "draft" sequence. Both sides agreed that it would not be a good idea for the public effort to appear to have been defeated, and that the announcement should be a joint one, a celebration of sorts, and should include a speeded-up timetable for the public effort.

## *Working Together*

So it was that on June 26, 2000, the first stage of the race that no one wanted to call a race was declared over. In a major presentation at the White House,

President Clinton appeared, along with Collins and Venter, and announced the development of not one but two working drafts of the entire human genome. The two groups, which clearly had been vying to be first and foremost in the race, had agreed to call it a tie and to concentrate on the accomplishment. "We are here," said the President, "to celebrate the completion of the first survey of the entire human genome. Without a doubt, this is the most important, most wondrous map ever produced by humankind."[39]

No one doubted that the payoff would be huge. As Venter has pointed out, for example, "The way medicine is practiced now, most drugs work on 30 to 60 percent of the population, yet we give the same drugs to the whole population."[40] In the future, a single blood sample may be enough to make prescriptions more effective and, with screening for variant genes that can cause adverse reactions to drugs, safer as well. Medical care in general will become more focused and less expensive.[41] Preventive medicine would become more than a slogan.

Nevertheless, it was, perhaps, a premature party. The HGP had not fully attained its target of a "rough draft" containing 90 percent of the total sequence. Venter, on the other hand, maintained that his sequence was 99 percent complete, but there were few who could verify his claim. Venter agreed not to point to the fact that he was ahead in numbers, and Collins agreed not to dwell on the fact that Celera had used the consortium's data, and that the consortium's draft sequence could be publicly searched for human genes, which was one of its stated objectives. Collins, in his remarks, was able to point out, "Already more than a dozen genes responsible for diseases from deafness to kidney disease to cancer have been identified using this resource just in the last year. So there is much to celebrate."[42]

Some thought it would be good for both results to be published at the same time and in the same major journal. That was not to be, and the result was another coordinated, but not joint, release of results. The HGP's draft sequence was published in *Nature* on February 15, 2001.[43] Venter's draft sequence data were published a day later in *Science* on February 16, 2001.[44]

All agreed, however, that this was not an end, but rather a beginning.[45] Even as the genes are found, functions must still be assigned to them. Some form of collaboration would help, but at the very least, both sets of results would play a role in the "finishing" process, which would include cleaning up the data, closing the gaps, and assigning functions to the genes.

Toward this end, said the President in his remarks, the two sides would work together. "[A]fter publication," for example, "both sets of teams will join together for an historic sequence analysis conference. Together they will examine what scientific insights have been gleaned from both efforts and how we can most judiciously proceed toward the next majestic horizons."[46]

In his remarks, Collins agreed and referred to "a continued, powerful, and dedicated partnership between basic science investigators in academia and their colleagues in the biotechnology and pharmaceutical industries."

Then he added, "And I wish to express my personal gratitude to Dr. Craig Venter for his openness in the cooperative planning process that led to this joint announcement. I am happy that today the only race we are talking about is the human race."[47]

Venter did not respond in kind, which, it would seem, spoke volumes. He did point out, however, "The beauty of science is that all important discoveries are made by building on the discoveries of others. I continue to be inspired by the work of the pioneering men and women in the broad array of disciplines that had been brought together to enable this great accomplishment."[48]

The detente lasted about half a year. Early in 2001, the wrangling started again. Lander and two other of the top HGP biologists made several stinging statements. Not only had Celera not beaten the HGP, they argued, but it had actually failed, in spite of having relied on the consortium's work to assemble its genome. Lander claimed, for example, that Celera's whole genome shotgun method "was a flop. No ifs, ands or buts. Celera did not independently produce a sequence of the genome at all. It rode piggyback on the consortium's efforts."[49]

Venter was not about to take this lying down: "We think there is zero legitimacy to anything Eric is saying, and we don't understand why he is saying it."[50]

But of course there were very good reasons for the HGP side's comments. Regardless of claims to the contrary on both sides, it was a race; and if there was a declared winner, that side would go down in history as the group that pulled off the greatest triumph in biological history. There were (and may still be) possible Nobel prizes. After HGP scientists had spent years laying the foundation, was Venter going to walk off with the honors?

Nor were the HGP scientists satisfied with an airing in the public press. The following year, they followed up with a major paper in the prestigious, peer-reviewed *Proceedings of the National Academy of Science*.[51] In addition to Lander, the article was signed by Robert H. Waterston, who headed the Genome Sequencing Center at Washington University/Saint Louis, and John E. Sulston, of the Wellcome Trust Genome Campus in Cambridge, England. "Our analysis," they wrote in the five-page paper, "indicates that the Celera paper provides neither a meaningful test of the WGS [whole genome shotgun] approach nor an independent sequence of the human genome." They were careful to add, "Our analysis does not imply that a WGS approach could not be successfully applied to assemble a draft sequence of a large mammalian genome, but merely that the Celera paper does not provide such evidence."[52] And to squelch any thoughts that the WGS approach could perhaps displace the consortium's method, they concluded: "Current evidence suggests that clone-based sequencing (the public group's method) will remain an essential aspect of producing finished sequence from large, complex genomes."[53]

The journal gave both sides a chance to comment on the paper in a following Commentary section. Phil Green of the Howard Hughes Medical Institute and the University of Washington, wrote for the prosecution that "accuracy tests in ref. 5 [the Venter paper in *Science*], which involved comparison of Celera's assembly to finished portions of the public sequence, are virtually meaningless because the finished sequence was itself used in constructing the Celera assembly." Also, "[W]hat Celera calls a whole-genome assembly was a failure by any reasonable standard: 20% of the genome is either missing altogether or is in the form of 116,000 small islands of sequence . . . that are unplaced, and for practical purposes unplaceable, on the genome."[54]

Referring to the Celera machines, he maintained that "when the public effort acquired similar capacity, they were able to attain a comparable or higher throughput by using the clone by clone approach."[55]

Writing for the defense were five of the major officers of Celera, including Venter. There is no room here to go into the details of their rebuttal, but in answer to the complaint that they used HGP data, they argued that they deliberately shredded the data into tiny pieces to eliminate any biases or errors, as well as positional information, that might have been present. The PNAS criticism was misguided, they said, because it oversimplified the mathematical procedure. Also, a simulation performed by WLS [the consortium authors] assumed that the Venter group had applied it to the entire genome, whereas, they say, it was applied to a single finished high-quality chromosome, constituting only 1 percent of the genome, with the result that the WLS analysis was "misleading."[56]

All in all, it seems clear that nothing is clear as yet.

What has seemed evident to many writers is that Venter's entry into the ring had, overall, a positive effect on the genome project. It did seem to speed things up, and provided two different approaches that could either be combined, or chosen for use depending on what the researcher is trying to accomplish.

But even here, there is far from complete agreement. Green wrote in his paper of several "myths." One myth, he says, is that the whole-genome approach was inherently faster and explained Celera's ability to churn out results faster than the HGP. Green says not so. The key factor, he maintains, was their use of "a huge, unprecedented squencing capacity . . . as a result of their corporate ties with a manufacturer of these machines. That this was really the key factor is evident from the fact that when the public effort acquired similar capacity, they were able to attain a comparable or higher throughput by using the clone by clone approach."[57]

Another myth, he wrote, is that the whole-genome approach saves money. When the finishing process is taken into account, however, "none of the costs that were supposed to be saved by the whole genome shotgun in fact were."[58]

Finally, he summarized, "A widespread view among many observers has been that . . . the genome race has in any case at least been good for science. In my view this is also a myth." He feels that although competition has some beneficial effects, "it also has the downside of encouraging shortcuts that may compromise the ultimate result. In the case of the genome race, the downside seems to have outweighed the benefits." Both Celera and the HGP, he wrote, took shortcuts that may hurt in the long run. For example, "it remains quite unclear whether the decision to produce an intermediate quality product (the draft) will prove wise in the long run; although the major centers have stated a commitment to finish the genome, motivation of many participants has surely been reduced now that the project is regarded by the public as complete. It remains to be seen whether a truly finished genome will appear by next year [2003] as promised."[59]

He then summarized what he called the undesirable results of the competition: "widespread misinformation, exaggerated claims, and a compromised product . . . mostly due to the high-profile nature of the contest, and perhaps also to the fact that a significant amount of corporate money was riding on the perceived success of one team."[60]

## *Irony*

This makes it seem as if Celera had profited, or stood to profit, enormously from Venter's audacious moves. Ironically, that is not what has happened at all. Though Venter tended to be the maverick, he was aware of the tightrope he was walking. The patent issue alone was clearly a significant one, enough so to bring in the leaders of the United States and Great Britain. On March 14, 2000, President Clinton and Prime Minister Tony Blair issued a joint statement on gene patenting. Blair was speaking for that country's public genome effort, which, via the Wellcome Trust and the Sanger Centre (the country's prime sequencing institute), made up a significant portion of the HGP. Their statement argued that data from gene mapping efforts should be public property and should be freely available to the public. Result: biotech stocks, including Celera's, plummeted. Clinton later amended his statement to suggest that there were reasonable commercial applications. Stocks recovered somewhat.

The rockiness of Celera's position became even more evident at the joint "rough draft" announcement on June 26, 2000. After President Clinton's celebratory comments, and in recognition of Celera's somewhat delicate position, he added, "I want to emphasize that biotechnology companies are absolutely essential in this endeavor. For it is they who will bring to the market the life-enhancing applications of the information from the human genome."[61] Nevertheless, on the day of the announcement, Celera's stock dropped 10 percent. In two trading days, it fell from $127 to $99.50. To investors, it apparently seemed that Celera was becoming entirely too chummy with the public project and that the commercial prospects would suffer. In fact, the more Venter assured the public of his openhandedness, the worse it seemed for Celera.

By the beginning of 2002, Celera's stock had dropped to $22, a far cry from the stock's recorded high of $275, reached a few months before the announcement of the genome draft.[62] Tony White, the chairman and chief executive of Applera Corporation, the holding company that created Celera, announced that it was time for Celera to make the move it had been planning for years, that is, into finding new drugs and clinical tests. And for that it needed a chief executive with different skills than those possessed by Venter. Result: Venter—the central scientific pillar of the company—resigned in January.

Not one to dwell on reversal, Venter announced in August of the same year his plans to build the nation's largest genome sequencing center. It will, he promised, introduce new techniques and technology that greatly decrease the time and effort needed for sequencing the code not only of humans, but of animals and microbes. He is providing the initial financing, using funds earned from Celera and his previous enterprises. He remains on the board of TIGR and has started two new smaller institutes as well, which will pursue various aspects of genomics.[63]

The new center will be a not-for-profit enterprise. As such it not only will be in competition with Celera for certain applications, but will also compete for federal funding, which would put it in competition with the university sequencing centers that have been working with the Human Genome Project.

Meanwhile the HGP continues its steady pace and is still working on a final version of the human genome, as well as working out plans for the finishing phases. The "final draft" was completed in late 2001. All work since then has been on finishing. By late September 2002, over 90 percent of the work had been finished. In April 2003, the HGP was able to report the successful conclusion of the project, as planned.[64]

## *A Marriage—of Sorts*

An editorial that accompanied the February 16, 2001, *Science* publication of the Venter rough draft stated:

> This stunning achievement has been portrayed—often unfairly—as a competition between two ventures, one public and one private. . . . In truth, each project contributed to the other. . . . Thus, we can salute what has become, in the end, not a contest but a marriage (perhaps encouraged by a shotgun) between public funding and private entrepeneurship.
>
> There are excellent scientific reasons for applauding an outcome that has given us two winners.[65]

Why not leave it at that?

# CHAPTER 10

# *Rifkin versus the Monsanto Company*

## Battling the Biotech World

> Our . . . biotech crops provide solutions for pest and weed control that can have added benefits for growers, consumers and the environment, including a reduction in the number of pesticide sprays and reduced environmental exposure, reduced labor, [and ] higher yields . . . while respecting the environment as well as regional and cultural diversity.[1]

> The new genetic commerce raises more troubling issues than any other economic revolution in history. . . . Will the mass release of thousands of genetically engineered life forms into the environment cause catastrophic genetic pollution and irreversible damage to the biosphere?[2]

The first quote appears on the Web site of the Monsanto Company, one of the world leaders in the development and marketing of genetically modified organisms (GMOs). The second comes from Jeremy Rifkin, an economic activist who has been fighting GMOs for a quarter of a century.

Using every activist tool possible—boycotts, lawsuits, demonstrations, personal appearances, newspaper ads, and interviews, plus an incredible outpouring of written material—Rifkin has taken on the emerging world of agricultural biogenetics. His opposition to the use of modern genetic techniques to improve various aspects of food production is legendary. He has in fact been labeled a modern Luddite.[3] David Baltimore, the Nobel prize–winning biochemist, calls Rifkin a "biological fundmentalist."[4]

Controversy seems to be his middle name. Kristina Shalizi, a staff member at the Santa Fe Institute, a think tank, calls him a "dangerous loon."[5] An Italian organization called Hypothesis: Access to People and Ideas feels he is worthy of a Nobel prize.[6]

While Rifkin's targets have included industry, government, and the scientific establishment, he has aimed his main artillery at the Monsanto

Company—his idea of a monopolistic monster that stands for everything evil and that represents for him the food biotech industry as a whole.

A quarter of a century ago, his was a voice crying in the wilderness. But his techniques, and his persistence, have paid off, big-time. He is seeing results—in the way of powerful and growing opposition—that would have been unimaginable even a decade ago.

His battle is by no means won, but the biotech giants, as well as much of the rest of the scientific establishment, have been shocked to discover that his efforts seem to be resonating in all sorts of surprising places.

Genetically modified food, he says, will become "the single greatest failure in the history of capitalism in introducing a new technology into the marketplace."[7]

Maybe, maybe not. Dean Bushey, the manager of genomics technology at Aventis CropScience, avers, "[E]ventually the genetically modified organisms are going to win, because it's such a powerful, such a beneficial technology."[8]

In the meantime, Rifkin is making lots of waves and now has plenty of company.

## *Beginnings*

In the 1970s, biological researchers saw a remarkable phenomenon: bacteria were swapping parts of their DNA with one another, and not just once in a while, but with great frequency. Then scientists figured out how to cut a gene out of one organism and insert it into another, with that gene becoming a part of the second organism's genome. Genetic engineering was on its way. The first important commercial results were in pharmaceuticals. For example, Genentech, a small, privately owned company, found a way to produce somatostatin, a hormone found in the human brain. Then it produced insulin that was actually safer than the standard, bovine-derived variety.

At the same time, fears about the dangers of genetic engineering erupted and led to a self-imposed moratorium by the scientific world on gene-splicing experiments until there was further evidence of its safety. As this evidence mounted, the fears diminished and the world of pharmaceutical biogenetics exploded into action.

Then a curious thing happened. It had been known for decades that cows given a natural product, bovine growth hormone (BGH), produced more milk. But it had to be derived from dead cows. With newly developed techniques, Genentech produced a purified, recombinant form, named rBGH, in 1981. Because it was produced under controlled laboratory conditions rather than derived from dead cows, the danger of disease transmission was lessened. It looked extremely promising, and Monsanto, brimming with enthusiasm, bought the rights to market it.

At the same time, there were organizations already in action that had been fighting for a variety of causes—against toxic wastes, in support of wilderness areas, in defense of the small farmer. Whatever their cause, genetic engineering seemed to them to be an unwanted intrusion on the natural world. The groups didn't like rBGH, but felt it was not a hot issue. By the mid-1980s, however, Rifkin had told National Wildlife Federation's Margaret Mellon, "I'll make it an issue! I'll find something! It's the first product of biotechnology out the door, and I'm going to fight it!"[9]

It's not clear just what he did to make it an issue. Perhaps it was just his declamation that got the environmental groups going. In any case, it became an issue. The anti-GMO groups, in fact, thought up all kinds of interesting possibilities, including cancer in humans, increased stress in the cows receiving the hormones, and more. A group of small dairy producers added to the criticism and won creation of laws in several dairy states that required labeling of milk that had come from such cows. Negotiating the various regulatory and critical hurdles that began to erupt took more than a decade. The FDA finally decided it did not present any threat to humans and approved it in 1993. Monsanto put it on the market the next year, and still sells it.

Even though no deleterious effects were ever seen as a result of drinking milk from cows that had been fed the new hormone, genetically engineered BGH milk turned out to have been a bad first product for Monsanto. It seemed to benefit the producer, especially the large producer, and offered little to the consumer. A Monsanto executive, Richard Mahoney, has stated, "It's come out all right, but if I'd known the public furor and the difficulty we'd face, I wouldn't have gone into it at all."[10]

End of story? Not quite. As of 1998, Rifkin was still warning that "the stress placed on the animals often leads to increased illness and suffering." He also speaks of "angry dairy farmers across the country, many of whom are reporting increased health problems in their herds after administering the new genetically engineered drug."[11]

Monsanto's Web site suggests otherwise and includes some testimonials from farmers who have been using the product. Melvin Brown Jr. of Milltown, Kentucky, states, for example, "Posilac has helped me to keep good cows in the herd longer. I can now keep them producing while waiting for problem breeders to freshen again."[12]

Whom should we believe?

As the milk hormone case continued to percolate, Monsanto and other biotech companies turned their attention to what they hoped would be a less contentious field.

## *Genetically Modified Foods*

Though the technology of genetic modification of plants is relatively new, the basic idea is not. Horticulturalists have been selectively breeding crops for thousands of years. Virtually all our crops have gone through this process. Using a variety of techniques, traditional breeders have the same objectives in mind: they try to develop varieties that are higher yielding, better adapted to different environments, more resistant to pests and diseases, easier to handle mechanically, and/or with improved quality and flavor.

Using such processes as taking a cutting from one plant that has certain characteristics and grafting it to a base plant, anywhere from a few to thousands of genes might be inserted into the base plant. When the treated plant matures and drops seeds, it is hoped these contain the additional desired characteristics. But the process is basically trial and error, with results not apparent until the new plant has grown.

There is another important difference between the two methods. Traditional breeding only works within the same species. With the new techniques of genetic modification, a few *selected* genes can be transplanted from a stunning array of other species—bacteria, viruses, other plants, fish, and animals.

In 1983, scientists at the University of California/Berkeley came up with a genetically modified bacterium, *Pseudomonas syringae*, that they could put into crop plants. They had found that by this means, modified plants could sustain lower temperatures (by a few degrees) before succumbing to frost. They wanted to test it on a larger basis by running a field trial on potatoes.

At this point, Rifkin had tasted blood and found he liked it. Thanks to him, the controversy evolved into the "ice-minus case." Rifkin found a weakness in federal regulations and sued the university. He was able to gather testimony by several well-known ecologists that the effects of the release of such bacteria into the environment could not be predicted and that what these scientists were planning to do could be dangerous. Although it is precisely for this reason that field trials are held, typically under carefully controlled conditions, the case escalated to the point where Congressman Al Gore held congressional hearings into the environmental effects of genetically modified crops.

At about the same time, a company called Advanced Genetic Systems wanted to carry out similar tests on strawberries. For five years, the proposal was kicked back and forth between federal regulators, who were generally in favor of it, and the courts, which called for delay. By 1987, the company was given permission to try out its bacteria by spraying an acre of strawberries. As a result of all the hullabaloo, the sprayers were encased in moon-type suits, and news photographers had a field day. Those pictures well supported the old adage, "A picture is worth a thousand words."

As a result of the lawsuit, the hearing, and the accompanying publicity, plus the fact that the FDA had labeled the bacterium a pesticide—which brought in the Environmental Protection Agency (EPA) because of its potential effect on the environment—the bacterium's developers simply gave up, and Advanced Genetic Systems ran out of money. Subsequent freezes in California in 1990 and 1998 cost citrus growers some $600 million to $800 million, some of which surely would have been saved if the method had had an easier road to development.[13]

Rifkin, naturally, sees things differently. In his book, *The Biotech Century* (1998), he continues to argue:

> The environmental impact statement [re: ice-minus] eventually was completed and the field experiment took place, despite the fact that there existed little in the way of a risk assessment science to judge the potential impact of releasing ice-minus . . . into the open environment. The government, the nation's molecular biologists, and the biotech companies continued the charade, contending that a sufficient body of science existed to measure "the risks," and that adequate regulatory safeguards were in place. . . . For the most part, the media and the scientific press were complicitous.[14]

The biotech companies obviously disagree. As Daniel Charles, who has been following this field for years, puts it: "The biotech industry has blamed those pictures ever since for cementing an image of danger in the public mind."[15]

Nevertheless, visions of taking advantageous traits (genes) from just about any organism and inserting them into crop plants produced extraordinary excitement in the worlds of genetics and agricultural science. The possibilities went far beyond frost resistance—better taste, longer-lasting produce, higher nutrition, salt-tolerant crops, even the transfer of nitrogen-fixing ability from beans to cereals, which would cut the need for chemical fertilizers way down.[16] It didn't take long for business to see the enormous promise.

But the very same developments, which seemed so promising to some, lit a different kind of fire under Rifkin, a fire that has found plenty of fuel over the last several decades.

## *Rifkin*

One author, Blaine Harden, describes Jeremy Rifkin as a "professional multi-issue radical." Harden continues: "He came to Washington [in 1969] as an anti-war activist and has made his living here since by skipping from issue to issue, protesting while an issue is hot and moving on when it is not."[17]

For over a quarter of a century, Rifkin has issued, on the average of one every couple of years, a book crying out against some aspect of new technology. One of his first targets was nuclear power. But it soon became clear that this his real mission in life would be fighting genetic engineering.

His training, however, was in economics and international affairs, and in 1977, he created the Foundation on Economic Trends. Nevertheless, it has been from there that he has launched his continuing attacks on the world of genetic modification. In the same year, he published a book called *Who Should Play God?: The Artificial Creation of Life and What It Means for the Future of the Human Race.* Although he wrote the book with a coauthor, Ted Howard—a former elementary school teacher and antiwar activist—I am assuming the following quotes reflect his own feelings.

Regarding in vitro fertilization for infertile couples, for example, he maintained:

> As a tool for uniting carefully selected sperm and eggs to produce some predetermined quality—"dial-a-baby," in the words of one observer—embryo implants and in vitro fertilization are a step closer to genetic engineering. The way is also being opened to the modification of life at the molecular level through the use of in vitro techniques. . . . Another giant step along that road is being taken right now as biologists strive to move beyond the womb and embryo implants into the realm of life grown completely in vitro—the true test-tube baby.[18]

This quote is interesting on several counts. He has found, and co-opted, someone else's clever phrases, "dial-a-baby" and "test-tube baby," and used them to good effect. Later on, he would popularize such terms as "Frankenstein foods" and "Frankenfoods."

Second, he has taken a promising technology and sought out the worst possible scenario that *might* arise from its use. From that fearful postulate, he concludes that genetic techniques should therefore not be used at all, regardless of how much relief of suffering and anguish they might provide to those in need.

Finally, it shows his early feelings about genetic engineering, feelings that apparently have not softened at all over the years, but have changed targets as each new example of capitalistic/industrial/technological activity has surfaced.

He has plenty of critics: Matthew Hoffman, of the Competitive Enterprise Institute in Washington, D.C., describes Rifkin as a "very unscrupulous man, a sophist, [and] a master media manipulator [who] loves to make all sorts of hysterical . . . predictions which the press picks up on. . . . If there is anything he hates, it is technology."[19]

Rifkin, of course, denies this. "People say I'm against the future," he argues. "I'm just opposed to the future they have in mind for us."[20] He doesn't specify who "they" refers to.

Rifkin set his tone early. In *Who Should Play God,* referring to the just-emerging techniques of human genetic engineering, he wrote, "This is a form of annihilation every bit as deadly as nuclear holocaust, and even more profound." Conjuring up the demoralized, depraved society described by Aldous Huxley in *Brave New World,* he added, "[W]hatever forms of future beings are developed will be forced to live the consequence of the biological designs that were molded for them." Implicitly recalling the hopes of Victorian eugenicists (ideas and desires long ago rejected by most responsible scientists), he also predicted "a mass program of biological engineering over the next twenty-five to fifty years."[21] Thus far his apocalyptic prediction of a mass program of human biological makeover does not seem to be happening.

Although he claims now that he has never opposed biotechnology for such applications as genetic screening, this position is a little hard to credit.[22] In the same book, he stated, "The rapid growth in the field of [genetic] screening is setting the climate for a revitalized negative eugenics, where 'defectives' (defined in whatever way society determines) will be disposed of like the unwanted babies of ancient Sparta."[23]

In 1998, he was still using the idea: "When Aldous Huxley wrote his dystopian novel *Brave New World* in 1932, neither he nor his contemporaries could have imagined that by the end of the twentieth century the scientific insights and technological know-how would be in place to make real his vision of a eugenic civilization."[24] In other words, he does not fault the technology per se from a scientific/technical point of view, but rather tries to picture the worst possible outcome(s), the potential for abuse, and goes on from there. But rather than see the abuse as something to be prevented, he wants the technology stopped, period.

Rifkin was not the only one who saw that a biological revolution was in progress, of course. Nor was he even first. Earlier, in 1963, Richard P. Feynman—iconoclast, polymath, and Nobel prize–winning physicist—warned, "[I]n the near future the developments in biology will make problems like no one has ever seen before. The very rapid developments of biology are going to cause all kinds of very exciting problems."[25]

But note the difference in tone, and specifically Feynman's use of the term *exciting problems*. He, like many other scientists, saw these problems as a challenge rather than as a sentence of doom.

In 1983, Rifkin wrote a book called *Algeny*. The word was his own creation, suggesting a combination of alchemy and genetics. Although the word never caught on, his warnings continued along similar lines: "While the nation has begun to turn its attention to the dangers of nuclear war, little or no debate has taken place over the emergence of an entirely new technology [genetic engineering] that in time could very well pose as serious a threat to the existence of life on this planet as the bomb itself."[26]

Arguing against human gene therapy, including work with bone-marrow cells, he maintained not only that "the scientists are rushing into this because of pressure from the marketplace," but also, "Suffering is part of every species' existence."[27]

In 1998, Rifkin was still announcing the end of the world: "Will the creation, mass production, and wholesale release of thousands of genetically engineered life forms into the environment cause irreversible damage to the biosphere, making genetic pollution an even greater threat to the planet than nuclear and petrochemical pollution?"[28]

## *The Promise*

Ironically, much of the early work with genetically modified plants was done by idealistic researchers driven by the hope of dealing in a better fashion with some of the ongoing agricultural and chemical mistreatment that our earth has been sustaining. Chemical pesticides, for example, had made large-scale farming both possible and efficient, but have clearly been wreaking havoc with the environment downstream—as depicted so graphically (though not necessarily always with complete accuracy) by Rachel Carson in *Silent Spring* (1962).

Many researchers entering the biotech field were looking for a way out of the chemical morass—even to the point of hoping to break the grip of powerful chemical companies, including DuPont, Dow Chemical, and Monsanto. They really saw themselves as saviors of the planet.

Their work, they hoped and believed, on several levels would help:

- the farmer, with more efficient techniques that would deal with such problems as weeds, insects, and frost

- the consumer, with better-tasting foods that stayed fresher longer
- third World peoples, by providing more, and more nutritious, foods
- the environment, by using fewer pesticides, and low- and even no-till agriculture; by inducing nitrogen fixation in cereal crops, to help decrease use of commercial fertilizers; and by reducing chemical runoff into our water supply

It seemed clear that this was a win-win situation. Everyone stood to gain. But as Daniel Charles suggests, "[T]his self-image held a hazard. Those who occupy, in their own minds, the moral high ground are usually the least able to accept criticism or even comprehend it. When the genetic engineers found themselves attacked by a new generation of environmentalists, they were incredulous and hostile."[29]

As the biotech field developed, yet another ironic transformation took place. Several of the major chemical companies, seeing the potential of agricultural biotechnology—and the possible threat to their chemical products—thought it would be wise to move into the new field via acquisitions, mergers, and divestment.

## *Monsanto*

Monsanto, for example, started life as a chemical company in 1901. Beginning with the production of saccharin, an artificial sweetener, it gradually diversified into plastics, food additives, and, especially, pesticides. By the 1990s, entranced by the promise of genetically modified foods, it decided to concentrate on this field. It both divested itself of the other product lines and bought up a variety of biotechnology-related companies. By the year 2000, Monsanto had 14,000 employees and generated $149 million in profits on $5.5 billion in revenue.

Other companies, including Dow Chemical and DuPont, began a similar process, either converting to, or creating, new biotech firms. This resulted in a new generation of GM opponents who saw themselves as challengers not only of genetic modification technology, but of powerful, entrenched big business interests. As Charles puts it: "The grander Monsanto's vision became, the more ominous the company appeared."[30]

Rifkin's magic was already at work in this area, as well. According to Rifkin and a broad range of opponents, almost everyone is complicit—not only big business, but also genetic and agricultural scientists and engineers in love with their own research and even government regulators.

## *Regulation*

"For the most part," Rifkin complains, "the scramble for fame and fortune corrupted the entire regulatory process, with government officials, corporate executives, and molecular biologists working side by side to assure the quick and expedient introduction of genetically modified organisms into the environment."[31]

He also finds fault with field tests—small-to-medium-scale testing that comes after laboratory work and is aimed at detecting potential problems before commercial release of the new product. GMO proponents, he says "contend that field tests, though inadequate, are better than no tests at all." He maintains, however, that "if the field tests are designed in such a way as to reveal little or nothing of the potential risk that might occur in large-scale commercial releases, then the exercise is little more than a regulatory farce, an elaborate fiction giving the appearance of scientific legitimacy without the substance."[32]

But as we saw earlier in this chapter, right from the very beginning (mid-1980s), GM researchers were already finding that they in fact could not march ahead unhindered. In 1986, Monsanto, too, came up against the Environmental Protection Agency and local authorities, thanks in part to objections raised by Rifkin. He questioned not only the safety of the tests on their gene-altered pesticide, but also the social implications of the work. As a result, some local strictures cut so deeply into Monsanto's plans that, even though they had already sunk $2 million into the experiment, they actually abandoned it when the growing season came and went while the EPA was still calling for further work.[33]

Recognizing the potential dangers of collusion, as alluded to by Rifkin, above, congressional legislation insists that EPA investigations include outside, independent scientists who have no financial stake in the outcome.

Currently three federal agencies share oversight of GM foods. In fact, says L. Val Giddings (the vice president for food and agriculture at the Biotechnology Industry Organization), "Crops and foods improved through biotechnology have been subjected to more scrutiny—in advance and in more depth and detail—than any foods in history."[34]

He explains:

> Three federal agencies share responsibility for ensuring, in advance, that foods and crops developed through biotechnology are at least as safe as their "conventional" counterparts. The U.S. Department of Agriculture (USDA), the Environmental Protection Agency (EPA), and the Food and Drug Admin-

> istration (FDA) oversee crops and foods according to a road map laid out in 1986—known as the "Coordinated Framework."
>
> According to this framework, the USDA handles any potential threats to U.S. agriculture and the environment, the FDA safeguards the food supply, and the EPA deals both with crops improved to resist insects and with environmental issues not covered elsewhere.[35]

"In the course of this regulatory oversight," Giddings continues, "the typical crop developed through biotechnology travels a research, development and regulatory pathway that takes, on average, 12 years and generates volumes of analytical and safety studies."

Mark L. Winston, a professor of Biological Sciences at Simon Fraser University in British Columbia, tends to agree. "This system of testing," he writes, "is thorough and effective, and it is difficult to argue on any rational scientific grounds that a GM crop passing the full battery of these tests should concern consumers."[36]

Still, there is another aspect of the regulatory process that worries GM opponents, and it goes by the name *substantial equivalence.* If a new product is similar to products already being sold and it has no constituents that are not already present in approved foods, then there is assumed to be substantial equivalence, and no further testing is needed. (The term has been criticized as too vague and confusing and is being abandoned in some quarters.) GM critics argue, nevertheless, that this is not a satisfactory reason for approving GM foods; there could, for example, be subtle differences that could cause trouble later on.

Until recently, the FDA felt that most GM foods are virtually equivalent to those they replace, and that no further tests are necessary; the EPA, dealing with the pesticide equation, has been less willing to give its assent. Starting in 2001, the FDA also tightened its requirements and began calling for specific tests on certain of the products.

## *The StarLink Story*

Another type of objection was voiced by GM opponents when regulatory agencies granted a temporary registration for a GM product, perhaps on the basis of substantial equivalence, before the safety testing process had been completed.

Regardless of how good, or how promising, a product appears to be, unless proper care is taken in its development and marketing, things can go seriously awry. A good example is the bacterium *Bacillus thuringiensis* (Bt), a soil-dwelling organism. Bt contains a gene that produces a toxic protein that is deadly to butterfly and moth caterpillars, some of which are serious crop pests. Developed as a biological insect treatment, it has no deleterious effects on humans or beneficial insects and has been widely used for decades for pest control. Bt is grown in fermentation vats and sprayed onto crops.

The system is certainly an improvement over chemical pesticides, which can harm beneficial insects and even humans. But there are some drawbacks: The Bt spray only works for a few hours, is expensive, and sometimes fails to penetrate the pest's system deeply enough to do its work.

Biotech researchers came up with the idea of inserting the appropriate bacterial genes directly into the genome of susceptible plants, including corn and cotton. Only pests that try to feed on the plants are affected. The farmer saves the cost of multiple sprayings, and the environment is spared a considerable amount of chemical pesticide use. The GM product was eventually sold under such names as Bollgard, NewLeaf, Nature-Gard, and StarLink.

Early on, a company called Aventis CropScience came up with a Bt variety of corn that was tested, along with other Bt products by other companies, in the mid-1990s. The products passed all the tests, with one small exception. The insect-killing protein used in Aventis's StarLink, Cry9C, failed one test for allergenicity: it survived for half an hour during a simulated test in gastric fluid, whereas it should have been digested immediately. This did not say that StarLink would be an allergenic problem, only that it might be. As a result, the agency licensed StarLink corn for animal feed, though not for any human food products.

The partial license turned out to be a major blunder. Farmers, intrigued by the new seed, planted millions of acres with it, but they had not been warned, at least not with sufficient urgency, that they should segregate the StarLink corn from corn destined for human use. Even if they had been well warned, in the hectic conditions of getting the product to market, it was probably inevitable that some of the StarLink product would have gotten mixed in with standard corn crops. It was shortly discovered in taco shells and then in other corn products. Though the percentage of GM corn planted was only a tiny percentage of the country's entire corn crop, the economic damage to the industry was enormous. First, there was a complicated and expensive recall, but unfortunately, not before some of it was eaten by humans.

It didn't take long before health complaints started to surface, and GMO opponents made much of the accidental release. For example, William Freese, the senior policy analyst for Friends of the Earth, states: "The Star-

Link investigation . . . is flawed by FDA's failure to investigate details about hundreds of corn-related allergy complaints reported to the food industry. Instead, the government has investigated only those people who directly called the FDA. Finally, the government's investigation fails to account for the extra susceptibility of children to allergies. To the best of our knowledge, only one child was tested."[37]

In the same article, Giddings answers: "Here is a genuine black eye for the industry. . . . The fact that a variety of corn not approved for human use entered the food supply . . . was inexcusable." But, he adds, "StarLink represents a rare failure of the regulatory system, not a threat to human health. It is a failure the industry united to pledge would not be repeated, even before the federal regulatory agencies moved to close the regulatory gate through which it had slipped. . . . It never should have happened, and it will not happen again."[38]

The health complaints nevertheless led to a serious investigation, including an advisory panel convened by the EPA in November 2000. Reports on twenty-eight individuals from across the country who had experienced allergic reactions, and for whom corn products were among the suspected cause, were carefully investigated. The studies were carried out by both the FDA and the Centers for Disease Control (CDC). Testers looked for antibodies to the suspected toxin, Cry9C, in blood serum from those experiencing allergic reactions. No antibodies were found, indicating that the patients were not allergic to Cry9C.[39] Further, when investigators checked out samples of the foods consumers were complaining about, virtually none contained the GM corn in question.[40] An additional irony is that genetic modification is likely to be the only way to *remove* this trait from plants. Hypoallergenic rice, peanuts, and other crops are being developed.

Nevertheless, it was a public relations disaster, with a deluge of media reports that left an impression of risk that was serious and impending rather than theoretical and remote. Giddings points out, "Although humans around the world . . . have eaten hundreds of millions of tons of food derived from crops improved through biotechnology, there still is not so much as a single, solitary sniffle or headache positively linked to their consumption."[41]

## *Knights in Shining Armor*

The GM world had taken a serious punch, but carried on. Weed control is another area of great interest to GM researchers. One of the most difficult of the farmer's jobs, it has long been done either by hand—a backbreaking job—or by use of such herbicides as Monsanto's product, Roundup.

In the GM approach, biotechnologists have genetically modified specific crop plants to be resistant to the pesticide. Using such crops, farmers need only spray once while the crop is growing, thereby reducing or even eliminating the need for preplanting and post-harvest sprays. Monsanto's Roundup Ready crops have been one of its most important products.

The idea that the small farmer does wonderful things for the environment is simply not borne out by experience. Charles, who grew up on a farm, maintains, "The single most environmentally destructive human activity on the planet is agriculture. Clearing and plowing land in order to grow crops (even following organic methods[42]) amounts to an ecological disaster visited annually upon most of our planet's surface. . . . Agriculture is largely unregulated. Farmers can plant what they want. . . . They can plow right up to the edges of creeks, causing soil erosion; they can overdose their land with fertilizer or agricultural chemicals, placing nearby streams or ground water at risk. They can plant the same crops year after year, depleting the soil of nutrients and risking infestations of destructive pests or epidemics of plant disease."[43]

For these reasons, and many more, GM advocates expected that when they finally began commercial production, they would find smooth sailing. And they got it, at first. The initial introductions of GM products were auspicious. By the mid-1990s GM crops and foods were well launched in the United States and, it appeared, overseas. By 2001 in the United States, 68 percent of the soybean crop was genetically modified; for cotton, the figure was 69 percent; and for corn, 26 percent. Because GM food crops are also used extensively in packaged foods, it's likely that as much as 70 percent of all packaged foods contains GM ingredients. Other than the United States, with about 88 million acres of enhanced crops, countries with substantial acreage include Argentina, 29 million acres; Canada, 8 million acres; and China, 3.7 million acres—an increase of about 19 percent over 2000.

In other words, it looked like GM foods would be a shoo-in. Rifkin had been crying wolf for years in the United States, but had not gained much of a following. A chance meeting in 1986 was to change all that.

## *Europe*

As Monsanto, as well as other firms in the field, looked overseas for other markets, they found the story there, especially in Europe, very different from that in the United States. Again, Rifkin played a major role. But this time he had a very effective partner.

In 1983, Benedict Haerlin, a German activist, spent time in jail for his part in some violent "revolutionary"activities. A year later, an ascendant Green Party placed him on its list of candidates for the European Parliament, and to his surprise, the twenty-seven-year-old Haerlin found himself in a new job as a European lawmaker.

During a visit to the United States in 1986, he met Rifkin and was impressed by his arguments. On his return to Europe, he tried to generate some action against GM foods, but had little success. In 1990, he left the Parliament and returned to Germany. In 1991, he joined Greenpeace, an in-your-face activist group, and initially coordinated the German's office's campaign against toxic chemicals and pesticides.

At first, Greenpeace had little interest in GM foods, but as GM products began appearing on European supermarket shelves in 1996 and 1997, consumers and, especially, activists began taking notice. Greenpeace, with Haerlin leading a group of about fifteen full-time campaigners, began organizing boycotts, generating publicity, and staging demonstrations not only in Germany but in England, Denmark, and other European nations.

At first their actions attracted little attention. But in 1998, they got a major break. Prince Charles, a staunch advocate of organic farming—he has his own organic farm, carefully tended by his own crew—published a strong attack on genetically modified foods in Britain's *Daily Telegraph*. In addition to the usual argument—organic farming is better for the environment—he added, "I happen to believe that this kind of genetic engineering takes mankind into realms that belong to God and God alone."[44]

With respect to the second argument, there is little that can be said. As for organic farming being better for the environment, John Stewart, writing in *The Scotsman,* points out that organic farming is actually worse for the environment because the lower yields mean more land is in cultivation and therefore less is left to nature. He also states that organic farmers do use pesticides, including, in some cases, "even nastier ones like virulent, toxic and environmentally devastating copper sulphate to control potato blight, a practice long abandoned by conventional farmers."

Stewart concludes: "Subsidising inefficiency, pandering to cranks and propping up privilege have contributed greatly to the mess we are in now and we should not compound it by subsidising the absurdity of organic production."[45]

Nevertheless, Prince Charles's article had an extraordinary effect. Any kind of celebrity support for a cause would have been useful. From a high member of the royalty, it was priceless. Further, the British were still reeling from serious outbreaks of mad cow disease and an *E. coli* infection. Not only did the prince's objections resonate with questions about the purity of foods, but it also played into the government's difficulties in dealing with the outbreaks of disease.

Further, no other Europeans have the kind of affinity that Britons seem to have for their land, and GM foods were seen to be an affront against nature. Paul McCartney seconded the feeling. The article's effects also penetrated the continent. The Germans, for example, have also had a long history of romanticism involving a deep attachment to "nature," which the new foods seemed to be threatening.

Objectors also began to snipe at Monsanto and other companies for sneaking GM foods in without their knowledge. Groups such as the Soil Association, an organization of organic farmers, Friends of the Earth, and dozens of smaller organizations began to join in.

Greenpeace upped the ante. Its members began showing up in ports across the continent and demonstrating against ships bringing in genetically modified soybeans, one of the important GM crops. By 1997, they had the support of 250,000 consumers, with many of them complaining to grocery chains. In the same year, an activist group in Ireland, Genetic Concern, ripped up a Monsanto test plot of sugar beet, an important cash crop in Ireland. In this case, Monsanto had done for sugar beets what it had done earlier, in a bigger way, for corn, soybeans, and potatoes in the United States, and that is to transfer genes for herbicide tolerance.

From 1997 on, there were four separate acts of sabotage on test plots in Ireland alone. In France, a group attacked a storehouse of GM seeds in southern France. Governments that initially had been willing to work with the GM firms, as in France, began to have second thoughts.

Nor was this sort of violence restricted to Europe. In 2001, a group of radical environmentalists set fire to buildings at a laboratory for urban horticulture at the University of Washington and at a poplar tree nursery in northwestern Oregon. Both organizations were working with genetic modification. There were attacks in other parts of the country, as well.[46]

Monsanto tried to enlist some of its competitors in a public relations and advertising campaign to counter the onslaught, especially in Europe, but were rebuffed. Some of the activist groups had begun to attack what they saw as Monsanto's arrogance, and it's possible that Monsanto's competitors were not unhappy that the ire seemed directed at that company.

But all the biotech companies appeared to be prone to similar behavior. Dan Glickman, the secretary of Agriculture at the time, commented, "Instead of carefully reading the marketplace, the biotech industry has been captive instead to a kind of tone-deaf technophilia."[47]

The stronger negative response in Europe probably has several causes. There are interesting differences between the United States and Europe, a few of which were mentioned in chapter 1—for instance, that Americans are less bound by tradition and so are more willing to accept new developments. While both populations are suspicious of private enterprise, particularly large companies, Americans are more likely to trust their government and its regulators, and to question environmental groups, while the reverse is true in Europe. Further, the United States has not had a major food scare in decades, certainly not of the order of the mad cow disease outbreak.

Nevertheless, resistance began to be felt in the United States, as well.

## *A Chastened Monsanto*

By the end of 1998, a battered Monsanto began to rethink its attitude and approach.[48] A decade ago, it would have been hard to imagine Jeremy Rifkin being invited to address a Monsanto meeting. Yet in June 1999, that is exactly what happened.

Trying to see into the future of GMO, Monsanto and an associated group asked a consultant, Ulrich Goluke, to help them see thirty years ahead. Goluke was a specialist in a form of future gazing called story building, or scenario creation, a technique that large organizations have been finding useful in trying to see ahead. The idea is to create several potential, but contrasting, visions of the future. Rifkin, of course, was happy to help them foresee potential disasters; Monsanto hoped that they might avoid them by preemptive efforts.

Rifkin said later that he was impressed by the ability of the attendees to put aside their assumptions, but that he didn't think the meeting signaled an end to the feud, or that it would help heal the mutual dislike and distrust.[49] He was certainly correct there.

Then, a couple of months later, the reverse scenario took place. Robert Shapiro, who had been CEO at Monsanto since 1995, was invited to address a European planning conference on GMOs. Speaking to the gathering by satellite hookup, Shapiro swallowed his pride and said he realized his company had "irritated and antagonized people" in its efforts to introduce genetically engineered crops and food to the continent. Also, "Our confidence in this technology and our enthusiasm for it has, I think, widely been seen—and understandably so—as condescension or, indeed, arrogance. We are now publicly committed to dialogue with people and groups who have a stake in this issue. We are listening."[50]

At the same time, however, Monsanto and six other companies did join forces to create an advocacy group, the Council for Biotechnology

Information, through which it orchestrated a powerful, $50 million media campaign, trying to convince consumers that their products are safe for them and for the environment as well.[51]

The timing may have been a problem, in that the campaign came so soon after the mad cow and *E. coli* disasters. In any case, it all may have been too late. Not that the GM genie can be pushed back into the bottle. But the genie's path is going to be a lot more rocky than anyone would have guessed a decade ago. Starting in the late 1990s, even Monsanto's stock price began to waver and fall for the first time in years.

## *Attitude*

Part of the problem may well have been Monsanto's attitude. Company officials and researchers undoubtedly believed that they were doing good. But how the company went about it was often seen as arrogance. Daniel Charles, who is by no means an apologist for Rifkin, states, "I've come to accept the legitimacy of patents on some genetic constructions—the Bt genes are one example—that were built through human ingenuity, using elements found in nature. Yet I, along with many others, was repelled by Monsanto's rush to claim that genetic property and squeeze as much revenue as possible from it. The company's business ambitions, more than any other single factor, brought forth the backlash against agricultural technology."[52]

What should they have done? Bill Lambrecht, a newsman and the author of *Dinner at the New Gene Cafe,* says, "Rather than demonizing Rifkin as Monsanto initially sought to do, industry officials might have been better advised to learn ways to more effectively respond to him and to prepare communities for the attacks on genetically engineered foods that surely would be waged."[53]

In any case, Monsanto's strong-arm tactics probably did end up hurting themselves. But, curiously, Rifkin's actions may be having a similar negative consequence. Gregory E. Pence, in his book, *Designer Food,* makes an interesting point: "[A]ctivists such as Rifkin carp that large conglomerates control more and more of our . . . biotechnology, but because of critics such as Rifkin, only large corporations have the resources to battle critics for decades to bring new products to market and, after all that, to pay damages from suits if anything goes wrong. When small companies try to do this, they can't meet payroll and may go bankrupt."[54] Advanced Genetic Systems, mentioned earlier, is an example.

In any case, Monsanto merged with Pharmacia, a pharmaceutical company, in March 2000. The new company took the name Pharmacia;

Monsanto became the agricultural division, with Hendrik Verfaillie as its president and CEO. It was quite a comedown. As Daniel Charles puts it: "Monsanto the world-devouring giant had shrunk into the chastened subsidiary of [another] company."[55] Then, less than two and a half years later (August 2002), it was spun off by Pharmacia.

Nor was Shapiro's mea culpa a one-time thing. Verfaillie, who remained as head of the firm after it was spun off, continued along the same line. Under the heading, "Our Commitments" on the company's Web site, he talked about "The New Monsanto Pledge—a five-point declaration that compels Monsanto to listen more, consider our actions and their impact broadly, and lead responsibly." He added, "I hope that over time it [the Pledge] will help transform our company from a science-focused company to a company that has its basis in science but is also committed to its social responsibilities—a company that is open, transparent, and beneficial to all its stakeholders."[56]

Verfaille resigned in December 2002 after Monsanto had posted a series of disappointing profit reports. Five months later Hugh Grant was appointed to replace him, the company's third chief executive in four years.

## *Today: Europe and the Third World*

In the United States today, GM and non-GM foods are, in general, mixed together and can be sold without any special treatment or labeling. In a few cases, a product may have a label that says, "No GMO used in this product," but this is not required by law—at least not yet. There are movements afoot that could change this. Examples include a November 5, 2002 Election Day referendum in Oregon on mandatory labeling.[57] Both GMO opponents and supporters, seeing the referendum as an important bellwether, went all out in their campaigns. Although the measure was soundly defeated [70 percent against, 30 percent for], labeling proponents vow that they will continue the battle and that they will ultimately prevail. The U.S. Department of Agriculture is considering a voluntary certification scheme on the part of manufacturers that could be used by them if necessary for their export trade.[58]

The problems begin to mount, however, with GM foods or seeds that are to be exported. This is especially true for Europe, where the situation is in considerable flux, with a variety of rulings either in place or under consideration, and with different attitudes in different countries. Strictures are strongest in such countries as England, France, and the Nordic countries; much less so in southern countries such as Spain and Portugal.

The European Union, attempting to control the situation, actually instituted a moratorium on GM foods, but has recently proposed rules that would require exporters of GM foods to segregate GM crops from nonmodified crops before shipment to all fifteen countries of the EU. The Union hopes that stricter rules would prompt the end of the moratorium and give consumers enough information to make up their own mind. In essence they are calling for mandatory separation and labeling. The United States is lobbying against the move, arguing that it would inflate costs unnecessarily in an industry where the two forms are routinely mixed—and with no evidence of problems because of it. With the level of resistance as high as it is, however, mandatory crop segregation is likely to be instituted.

Fears of GM foods in some groups have escalated to a kind of hysteria. In Africa, for example, some 12 million people in six countries face starvation due to drought and plant disease, and the United States has shipped tons of GM maize (corn) to the region. While Malawi, Lesotho, and Swaziland have accepted the GM aid, Zambia, Zimbabwe, and Mozambique initially rejected it. The latter two finally agreed to accept the aid with the proviso that it be milled first to eliminate any possibility that seeds from GM plants would cross-pollinate with its natural plants.

Zambia, at the time of this writing (October 2002), has continued to hold out, with President Mwanawasa calling such food "poison," and arguing that it is intrinsically dangerous."[59]

Yet, Dr. Wim Van Eck, a food safety and nutrition expert at the UN World Health Organization, points out that a series of consultations with people at the Food and Agriculture Organization and the World Food Programme, as well as several national risk assessments, did not reveal any GM risks to humans.

And a United Nations Development Programme report released in 2001 concluded that GM technology holds great potential for helping developing nations feed their people. GM opponents, on the other hand, insist that the use of GM techniques in developing nations is wrong-headed, that in every country there is plenty of food to go around. The problem, they say, is that the people are too poor to buy the food that is available and so it is exported.

Some GM opponents feel even more strongly about the use of GM foods in the Third World. Robert Vint, national coordinator for Genetic Food Alert, says baldly, "Deliberate contamination through food aid neatly complements America's strategy of forcing GM food down the throats of the starving." He calls this an "act of diplomatic terrorism."[60]

On the other side, Ann Veneman, the U.S. secretary of Agriculture, argues, "Our ability to deliver desperately needed food has been greatly hindered by individuals and organizations that are opposed to biotechnology and who are providing misguided statements about the U.S. food system."[61]

Ironically, her statement was reprinted in a pro-labeling newsletter, the real objective being to set her up for attack: "Veneman's impartiality is questionable. . . . Before becoming Secretary of Agriculture, she served on the board of Calgene, Inc., which in 1994 was the first company to bring genetically engineered food, the Flavr Savr tomato, to supermarkets."[62]

And so it goes. In the meantime, as noted above, Europeans are calling for mandatory labeling of GM crops, and the FDA is thinking about calling for voluntary labeling instead. In Canada, a scientific panel has decided that the principle of substantial equivalence is not a satisfactory one, and that its regulators should do more in the way of specific testing for GM products. Winston agrees with this thinking; he feels "the old standard of substantial equivalence should be discarded, at least until consumer confidence is established. Few GM foods would fail a rigorous battery of tests, and the relatively small expense and slight delay in getting the products to market would be well worth the gain in consumer confidence."[63]

Bruce M. Chassy, a food safety expert at the University of Illinois/Urbana, disagrees. He maintains that "all GM foods are subjected to as thorough a battery of tests as we know how to do. . . . There is no indicator of unusual properties that would imply risks worthy of further evaluation."[64]

Meanwhile, Monsanto continues to bring out new products, such as a GM corn seed, whose plant will kill rootworm, a particularly voracious pest. Monsanto has high hopes for the product, which the EPA has found acceptable. A British analyst, John Roberts, says however that until Europe eases restrictions on GM foods, U.S. farmers will be slow to take up such products.[65]

## *Labeling, or Why Consumers Are Confused*

"Numerous polls," says Freese, "show that 80 to more than 90 percent of Americans want genetically engineered foods labeled as such. . . . The European Union (EU) and eight other nations already require labeling of genetically engineered foods. . . .

"If these novel foods really are as safe and beneficial as [proponents claim]," he continues, "why does the industry object to this simple measure? Why the secrecy? One possible reason is fear of liability. A growing number of doctors support labeling to permit assessment of whether their patients have experienced adverse health effects because of engineered foods."[66]

At every turn, one reads what appear to be diametrically opposed statements. Giddings answers that

> advocates of labeling for biotech foods are not calling for labels that give information about health, safety and nutrition that is accurate and informative. Such labels are required by law, have been supported by the biotech industry, and have included foods improved through biotech since 1992.
>
> Those clamoring for special labels simply want to stigmatize foods improved through biotech, while they misrepresent the rigorous regulatory process and proven safety of these products. Such labels will allow them to organize boycotts and economic blackmail.
>
> They argue that special labels are required to ensure "consumer choice" for those who want to stick to "natural" foods. Never mind that ten millennia of artificial selection by humans means that virtually nothing on a dinner plate in the industrial world would be recognized by our hunter-gatherer societies of not so long ago. . . .
>
> "Natural" corn has an ear the size of the last digit on your little finger. "Natural" tomatoes are the size of a grape, and a stunted one at that.[67]

Nor is it only the big corporations that are fighting in favor of this technology. In a test field near the town of Lerida, Spain, the journalist Martin Roberts reports seeing conventional corn plants that had collapsed in a recent storm. Many of the plants, riddled with pests, had rotten cobs. Nearby Bt plants were green and healthy. He quotes Fransisco Armengol, who farms one hundred acres of corn plants in the same region: "The new maize grows better and is more reliable; the old varieties used to fall down in the wind. Pesticides never used to work; we didn't use them."[68]

## *What Are We to Think?*

Rifkin has had a powerful effect on people and on the GM movement. Among scientists and intellectuals, however, he is often accused of poor scholarship and exaggeration. The late Stephen Jay Gould wrote a review of Howard's and Rifkin's *Algeny,* and called it "a cleverly constructed tract of anti-intellectual propaganda masquerading as scholarship. Among books promoted as serious intellectual statements by important thinkers, I don't think I have ever read a shoddier work."[69]

Gould objected, for example, to Rifkin's depiction of Darwin's evolutionary theory as a creation of industrial capitalism. He also pointed out:

"Rifkin does not respect the procedures of fair argument. He uses every debater's trick in the book to mischaracterize and trivialize the opposition, and to place his own dubious claims in a rosy light." Finally, "Few campaigns are more dangerous than emotional calls for proscription rather than thought."[70]

With objections like this, and with so many major organizations—Food and Drug Administration, Environmental Protection Agency, World Health Organization, American College of Nutrition, American Council on Science and Health, and many others—maintaining that GM foods are both helpful and safe, how have Rifkin and company been so successful? There are several parts to the answer.

## *A Powerful Mix*

First, Rifkin is very smart, very persistent, and very inventive. In 1992, ever on the lookout for new ways to make his point, he organized a boycott of GM foods by some well-known chefs.[71] In 1997, he and a colleague, the cellular biologist Stuart Newman, submitted an application for a patent on a human chimera. One definition of a chimera is an illusion or fabrication of the mind; another—and the one he certainly had in mind—is an imaginary monster compounded of incongruous parts.

To obtain a patent, one needn't actually create the idea or object, and the applicants apparently had no intention of attempting to create one.They only had to show that it is feasible. Actually, they were hoping for some sort of rejection on moral grounds, which they could then have used to fight GM patents on similar grounds. The Patent Office did reject the application, and two additional submissions, but for other reasons. Nevertheless, the concept, and its implications, got plenty of press.[72]

Two years later, in 1999, he gathered a group of antitrust lawyers from ten firms and filed a suit contending that Monsanto gave farmers false and fraudulent guarantees about the safety and marketability of a new breed of bioengineered seeds. Using terms reminiscent of the U.S. Justice Department's case against Microsoft, they argued that Monsanto has created a monopoly situation by bullying small farmers, by inimidation, and by using deceptive business practices.[73] They named nine other "co-conspiritors."

Whether Rifkin ever wins any of these suits seems to be quite irrelevant. Monsanto lawyers said right away that the suit was a political stunt, not a real legal challenge. Subsequent events seem to have borne this out. For example, a quick search on the Internet almost three years later brings up half a dozen articles on the institution of the suit, but not a single article describing any further activity.[74] Rifkin apparently feels that the publicity he got just from the filing made it worthwhile.

A *Wall Street Journal* article, however, concludes: "If some 16 years after he sued his first potato field Mr. Rifkin still worries about what you get when you cross a firefly with a plant, he's free to do so.[75] For our part, we're more worried about what you get when you cross a scaremonger with a tort lawyer."[76] Thomas G. Donlan, an editor at the financial journal *Barron's*, compared the suit to "A Visit From a Mob," and referred to the lawyers as "hired muscle."[77]

Another reason for Rifkin's success: he writes well. But as Pence points out, there's more to this point: "Thinking simplistically allows one to write with passion, without qualification, and with a style that suggests one is battling Absolute Evil. As readers of Jeremy Rifkin . . . also know, it is easier . . . to be eloquent when defending the status quo against technological change, for fear of such change taps into the Accident Story, our most ancient myth about the evils that befall humans who reach out too far."[78]

The press has also been helpful to Rifkin's cause, even aside from his many clever ways of generating publicity. In several cases—for example, one involving the monarch butterfly and one involving GM potatoes—initial reports seemed to provide scientific support for the anti-GM cause and got wide publicity. Environmental groups as well as anti-GM forces jumped on the reports.

On June 7, 1999, for example, Rifkin wrote in the *Boston Globe*: "On May 20, the term 'genetic pollution' officially entered the public lexicon. Scientists at Cornell University reported in the journal *Nature* that the pollen from genetically engineered corn containing a toxin gene killed 44 percent of the monarch butterfly caterpillars who fed on milkweed leaves dusted with it."[79] Shortly afterward, a bipartisan group of twenty congressmen introduced legislation calling for labeling of GM foods because of concerns over food safety.[80] By the end of the year almost a thousand newspaper and magazine articles had spread the word; the monarch had become the Bambi of the insect world.[81]

Later, more careful research showed the initial report to be flawed. In the fall of 2001, the National Academy of Sciences released the results of a multiorganization study that concluded: "This two-year study suggests that the impact of Bt corn pollen from current commercial hybrids on monarch butterfly populations is negligible."[82] But by then, who cared? It had nothing like the kick that the first, independent study had had, and it received much less play in the news.

A similar story played out with GM potatoes when Dr. Arpad Pusztai and a colleague published an article in the *Lancet,* one of the top medical journals, which suggested that raw GM potatoes appeared to thicken the lining of rat intestines.[83] Again, the anti-GM forces had a heyday. Again, subsequent evaluations severely criticized the article.[84] Again, the corrections got much less play in the news.

One wonders, then, why the *Lancet* chose to publish a flawed article. An analysis in the *Economist* states, "At least two of the paper's six reviewers recommended its rejection, enough to send most other manuscripts into the bin. But genetic modification is so fraught with public accusations of corporate cover-ups and government conspiracies that the *Lancet* decided it was right to publish."[85]

This of course plays right into the hands of the anti-GM forces, for, regardless of whether the original study was valid or not, its very publication in a respected journal adds powerful ammunition to the many weapons being mustered.

## *Risk*

A final reason that Rifkin's side is doing so well has to do with the subject of risk. Thanks to the fact that the concept is widely misunderstood by the general public, Rifkin can play easily with this idea, and he can take up a position that is difficult to counter. In a 1984 interview for *U.S. News & World Report,* he stated, "I would like to see the best scientific talent in the country come together to see if methods can be devised to minimize risks to the ecosystem. If they can't come up with such standards, it would be foolhardy to continue genetic engineering research."[86]

His position sounds reasonable at first. But in reality he's calling for a moratorium until all risks can be eliminated—which is impossible. In 1998, he was still asking, "Will the mass release of genetically engineered organisms in our biosphere mean genetic pollution and irreversible damage to the biosphere?"[87]

This is the basic idea behind what has come to be called the "precautionary principle," namely, that we should not allow a new technology or product to be introduced until it has been thoroughly tested and shown not to be harmful.

Thanks largely to the efforts of GM objectors, the precautionary principle has been incorporated into a number of international rulings and treaties, including the United Nations Biosafety Protocol. The problem for GM advocates is that the precautionary principle presumes that there are risk-free alternatives, which is hardly the case. There are plenty of "natural" foods, for example, that have poisoned or sickened consumers.

Peanuts, for instance, are terribly dangerous to some, and bacteria-induced food poisoning is a recurring problem. Certain food supplements are known to cause problems such as liver damage, stroke, seizures, heart attacks—and death. The FDA has received reports of a hundred deaths related to the herbal food supplement ephedra, also called ma huang, resulting in little action. The death of a young baseball pitcher in the spring of 2003 has brought some strong calls for regulation. Yet ephedra, or products very like it, still remains on the market. In fact, says the Society for Women's Health Research, "there is no widespread campaign to ban [herbal products], or even to require prior testing for safety and/or efficiency, as there has been for GM foods—for which there is not one single recorded health problem, in spite of the fact that millions of people have been eating such foods for years."[88]

A recent report reminds us that cosmetics can also be troublesome. It points out that many cosmetics contain phthalates—a class of chemicals known to cause developmental deformities in animals. The report adds that by claiming such substances are fragrances or trade secrets, manufacturers can legally keep them off ingredient lists.[89] A letter to *Science News* describes cosmetics as "a chemical stew with ingredients that would require EPA warnings if used in industrial settings."[90] The FDA recently reported that one hundred people die and more than two thousand are hospitalized each year as a result of liver damage from unintentional overdoses of acetaminophen (e.g., Tylenol, Percodan, and others).[91]

But the most surprising example is that of organic farming, touted by Prince Charles and other anti-GM advocates. The food expert Carl Winter points out, "Consumers need to understand that organic production does not mean pesticide-free and pathogen-free production." In fact, he adds, manure is a major source of fertilizer for organic farming. Although it is typically composted to kill bacteria, no composting is perfect, with the result that "health risks associated with disease-causing organisms [in organic farming] are far greater than risks associated with pesticide residues, which are negligible."[92]

Clearly, we do not live in a risk-free world. Yet the precautionary principle is essentially calling upon GM producers to prove a negative—that GM foods are risk free, which is virtually impossible to prove. Martin Teitel, director of an activist group called the Council for Responsible Genetics, admits that requiring GM producers to do so means essentially that "they don't get to do it, period."[93]

Actually, as Winston points out, "The gulf between regulators and the environmenal movement has widened, because the regulators focus on balancing benefit and risk while the environmentalists regard *any* risks as unacceptable."[94] [My italics.]

## *A Strange Campaign*

The anti-GM movement, which has taken on a life of its own, is replete with paradoxes. Bernard Dixon, writing in the *British Medical Journal*, summarizes them:

> Firstly, alarming headlines contrast with the fact that genetic engineering is regulated more keenly and transparently than any other technology in the past. Secondly, activists are destroying field trials specifically established to help in assessing the environmental impact of transgenic plants. Thirdly, critics talk of the imprecision of the new plant breeding methods. Yet these involve the transfer of one or at most a few genes—not entire chunks of genomes, as in traditional plant breeding and contrast starkly with the randomness of nature itself. Fourthly, the engineering of disease resistant plants, though opposed by some environmentalists today, is an example of the very approach that the pioneer environmentalist Rachael Carson advocated as preferable to the use of chemicals.[95]

He adds:

> [T]he row over genetically modified foods also contrasts oddly with the widespread acceptance and use of many recombinant products in health care. These include human insulin and growth hormone, erythropoietin, hepatitis B vaccine, tissue plasminogen activator, several interferons, factor VnL, and antihaemophilic factor. As pointed out by . . . focus group research in the United Kingdom, many people welcome medical applications of gene technology as "good genetics" but see genetically modified foods as "bad genetics."[96]

GM opponents argue that the problems with GM foods lie not only in its effect on human health but also in its potentially deleterious effects on the environment. "This seems irrational to many plant scientists," writes Robert C. Cowen in the *Christian Science Monitor*. "They cite statements by the National Academy of Sciences and by 11 scientific societies that insist there is no scientific difference between plants transformed by traditional methods and those transformed by transgene engineering. The issues of safety and acceptability should be decided on a plant by plant basis, not by the method that produced the new plant, these organizations say."[97]

Dixon adds,

> One obvious possibility [for trouble] is the transfer of genes from genetically modified crops to non-modified crops, where they may have unforeseen consequences. Yet, as observed in [a recent British Royal Society] report, well established practices already minimise such dangers in the case of conventional cultivars. For example, oilseed rape for human consumption contains low levels of erucic acid, which is toxic to humans, whereas industrial oilseed rape contains high levels. So the two varieties are grown sufficiently far apart to prevent cross pollination.[98]

As a result of what GM advocates see as irrational thinking, some of them have made some pretty strong statements, as well, and a suspicion has grown among some of them that the motives of the anti-GM forces may not be wholly idealistic. Just one example, which comes to us via Thomas DeGregori, of the Department of Economics, University of Houston: "If anything clearly emerges from this debate, it is that, when the veneer of pious rhetoric is stripped from the anti-GM food claims, their argument is simply one of selfishly seeking to impose their own fetishes and New Age beliefs on society. . . . It is really questionable whether anyone benefits . . . except those organizations that gain membership, funding, and above all, power from these Luddite actions."[99]

Charles remains hopeful: "In short, the chaos of public argument might actually yield an outcome that is both reasonable and wise."[100]

## *A Better Rifkin?*

As for Rifkin the man, he brings up a huge variety of responses, some of which we have already seen. One of the most interesting comes from one of his own supporters, David van Biema. Questioning some of the Rifkin's methods, he suggests, "It is easy enough to approve of the effects of Rifkin's work—biotechnology should be debated and regulated . . . Jeremy himself is engaging, fun company. But he is also an alarmist and an absolutist, with little or no trust in humans to think for themselves. One shudders for a world in which Rifkin is king.

"What one wishes for, ultimately, is a better Rifkin. Someone who does what he does but gets all his facts straight, honors the other side even while disagreeing with it, someone who has more faith in his fellow man.

"But," he added, "nowhere is it written that we always get the perfect Rifkin for the job."[101]

## *Alternative Biotechnologies*

Rifkin suggests in his latest book that "the genetic engineers might eventually lose their dominant position to the ecologists whose thinking is more in tune with a biosphere consciousness. If that were to happen, alternative biotechnologies might yet triumph over gene-splicing techniques in the Biotech Century."[102]

But future biotechnologies are not likely to be less powerful than today's, and as Rifkin himself puts it: "The more powerful the technology is at expropriating and controlling the forces of nature, the more exacting the price we will be forced to pay in terms of disruption and destruction wreaked on the ecosystems and social systems that sustain life."[103]

There is a possible alternative. Biotechnology's growing knowledge about the genomes of living things—including those of plants—suggests using this knowledge to enhance traditional breeding techniques. Once scientists know what genes are in a certain plant, they may be better able to select which plants to cross.

Although this gene-marker-assisted method is still a lot less efficient than genetic modification, it might avoid the regulatory requirements and the bad name that GMOs seem to have acquired. There is, on the other hand, still the possibility that once this marker-assisted method develops further, it might still run into the same sort of opposition as being against "nature."

So, if history is any guide, and if Rifkin is still around, it seems likely that he will by that time be arguing just as strongly against the new biotechnologies.

At a recent "Teach-In on Technology and Globalization," Rifkin declaimed, "This is the big wrestling match of the twenty-first century." "For once," says the science writer and editor Ronald Bailey, "he is indisputably correct."[104]

# EPILOGUE

Fear of technology has a long history. Even writing was at one time thought of with suspicion. In Plato's dialogue *Phaedrus,* for example, Socrates tells Phaedrus a story about the Egyptian god Theuth, who [says Plato] invented writing.[1] Theuth demonstrates it before King Thamus, and speaks of the great things it will accomplish: it will give the Egyptians better memories and make them wiser.

But Thamus suggests that the inventor may not be the best judge of the utility of an invention. He argues that writing would accomplish just the opposite of what Theuth claims. It would foster forgetfulness because the users will have less use for memory; they will rely on writing for ideas instead of relying on their own resources; they will receive information without proper instruction and will in consequence be thought knowledgeable when they are actually quite ignorant. And because they are filled with the conceit of wisdom instead of real wisdom, they will be tiresome company. Further, if a written work "gets into difficulties . . . it always needs its father to stand up for it."[2]

This is one of the earliest examples, and a remarkably prescient one, of what is sometimes referred to as the law of unintended consequences. But the "law" remained a sort of occasional, generalized fear until the turn of the seventeenth century, when the lawyer/man-of-affairs/philosopher Francis Bacon suggested that science and technology could be combined and used to master nature.

He, and many who followed, thought that mastering nature was a fine idea.[3] Others have been less sure, sometimes even seeing God frowning at the very idea, and fearing that overstepping these bounds will bring retribution later on.

Modern science and technology are certainly far more advanced and powerful than anything Bacon had in mind. During the last two centuries, the period covered in this book, the technologies have, in general, gotten progressively more complex. Have they created proportionately greater problems, problems that are outrunning society's ability to deal with them? Are the risks of modern technology worse than the benefits they offer? Arguments pro and con have gone on for years.

Yet, as you may have noticed, of all the controversies in this book, only the last one (Jeremy Rifkin versus Monsanto) actually deals with this important question. Elmo Zumwalt, in chapter 8, argued against nuclear propulsion, but more for economic reasons than because of any fear of its risks. And the Luddites, remember, were not technophobes in the modern sense.

That the subject of technophobia should not arise until the last chapter is not a coincidence. Public feelings about science and technology have been evolving in a way that made this almost inevitable.

The feeling after Bacon and during the Industrial Revolution was mostly in favor of new developments in science and technology, in spite of the occasional technophobic outbursts. The Victorian idea of progress held an optimistic view of science and technology that assumed society would keep moving toward a better place.

In our own era, the period after World War II saw strong public support for new developments in science and technology. Sputnik (1957) and, a dozen years later, the American moon walk, sparked a remarkable pro-science and pro-technology kick. There was a strong drive for increased literacy in both areas. C. P. Snow suggested in 1959 that one could not really be considered literate without having an understanding of basic scientific laws.

Since then, however, there's been a strange reversal. Not only is Snow's admonition passé, but a strong scholarly movement called postmodernism has even suggested that there is no such thing as objective science, and therefore that one need not know any science at all to evaluate it. American education seems to be reflecting this. Dorothy Nelkin, a New York University professor who has written extensively on science and society, reported that in spite of activists' demand for greater public accountability, "only about five percent of American adults are both attentive to science policy issues and sufficiently literate scientifically to understand and assess the arguments."[4] Visit any bookstore and see how many shelves are devoted to science and technology, and how many to mysticism, astrology, and the like.

Americans' distrust of industrial power has mushroomed, perhaps because recent revelations about the underhanded methods of many corporate leaders are only the latest in a variety of corporate problems and misdeeds. In 1973, Eugene S. Ferguson, a historian at the University of Delaware, considered the question of whether and how all segments of society should participate in deciding how technology might be used to solve contemporary problems. Ferguson maintained: "The pattern thus far, which has made the public wary of 'participation,' has been to seek public approval of projects that are already and irrevocably under way. Citizen groups have found that the technical community will listen to them only in a courtroom."[5]

Clearly, as chapter 10 has demonstrated, this is no longer the case. Although public feelings about technology have remained generally positive over the last quarter century, there has certainly been more questioning, and powerful interest groups have been exerting increasing pressure on a variety of technologies.

Result: technological disputes, too, are changing. As I searched the history of technology for dramatic examples of controversy, the ones that leaped out at me had little to do with fear of technology—until the last one.

"Controversies," Nelkin wrote, "are increasingly expressing moral judgments as well as economic interests, and they are becoming crusades."[6] This seems to fit the crusade against GMOs. Even Rifkin's use of the patent system was specifically aimed at advancing his cause, and had nothing to do with economic gain.

Even more important, the anti-GM group has, after all, accomplished much of what they set out to do, yet there seems little letup in their efforts.

Ironically, these powerful anti forces make less sense now than in the time of the Luddites. After all, the neo-Luddites are not fighting for their way of life—indeed for their very lives—as their predecessors were, but for their belief that the new technology will surely have apocalyptic results.

A further irony is that advanced technology is *more* likely to be able to foresee potential risks.[7] On the other hand, increasing emphasis on risk probably explains why there is more fear in the air than ever before.

Finally, only in an industrialized, technologically advanced society do people have the time, energy, and resources to mount a major, long-running campaign against a new technology, one that can last a quarter of a century and still be going full tilt.

Yet, we know that there are problems, and that the law of unintended consequences does emerge fairly often. Other writers have pointed out the perils—both real and imagined—of technology, sometimes eloquently, sometimes sensibly, sometimes wildly.

Among the best-known, and most eloquent, has been Jacques Ellul. For example:

> The reality is that man no longer has any means with which to subjugate technique,[8] which is not an intellectual, or even, as some would have it, a spiritual phenomenon. It is above all a sociological phenomenon. . . . But everything of a sociological character has had its character changed by technique. There is, therefore, nothing of a sociological character available to restrain technique, because everything in society is its servant. . . . Modern man divines that there is only one reasonable way out: to submit and take what profit he can. . . . If he is of a mind to oppose it, he finds himself really alone.[9]

Ellul wrote that in 1967. While it still provides food for thought, there is a curious postscript to his claims. For he also wrote: "[N]o one will publish a book attacking the real religion of our times, by which I mean the dominant social forces of the technological society."[10] It is ironic that not only was his book published, but it has become a classic, a must-read for anyone interested in the history or sociology of technology.

Certainly, as Ellul and others have pointed out, there have been plenty of downsides to the introduction of new technologies; and certainly there have been major dislocations at the time of their introduction (as was the case with the original Luddites). I still feel, however, that we can put the basic question as follows. Has modern technology (however you want to define it, but including genetically modified foods) made life better or worse for the average person?

This question does not have a simple answer. But a recent report from the National Center for Health Statistics states that the infant mortality rate in the United States has dropped to a record low—6.9 deaths per 1,000 live births—and life expectancy has hit a record high of 76.9 years. Compare this with another recently released figure: A wide-ranging study of health in the Western Hemisphere over the past seven thousand years concludes that in pre-Columbian hunter-gatherer societies in the Americas, few people survived past age fifty. Even in the healthiest cultures in the thousand years before Columbus, a life span of perhaps thirty-five years was the norm.[11] It would appear that modern science, technology, and medicine are doing something right.

"As much as the neo-Luddites might wish it otherwise," writes the bioethicist Gregory E. Pence, "there simply is no other social and economic model of lifting hundreds of millions of people out of poverty than what might be called democratic, technological capitalism."[12]

Yet, as we have seen, there is more suspicion of new technology as well as more insistence on public participation in its use and application. Perhaps, in the typical pendulum fashion of public feeling, scientific/technical literacy will again become fashionable, leading to the happy confluence of public participation with a greater public ability to evaluate the realities of benefit and risk.

Clearly, we're still finding our way.

# NOTES

## Introduction

1. Although the Luddites were going after a specific kind of manufacturing machinery, and are therefore commonly referred to as machine-breakers, the term *technology* actually has a much wider meaning. It generally refers to any tools, methods, or techniques in use for getting a job done, and so can include linguistic and intellectual tools, as well. The term first appeared in the English language in 1615. Modern Luddites are sometimes referred to as neo-Luddites.

2. Full quote: "The success of the Green Revolution was simply beyond question for all but those whose brains were eaten away by a Luddite ideology." DeGregori, 2001, p. 2. (Citation under Rifkin, Chapter 10.)

## Chapter 1. Ned Ludd versus the Industrial Revolution

1. Watson, 1993, p. 141.
2. Rossney, 1994, p. 2.
3. Sale, 1995, p. 4, footnote.
4. Thompson, 1966, p. 556.
5. Sale, 1995, p. 146.
6. Liversidge, 1972, p. 14.
7. Englander and Downing, 1988, unpaged, obtained online.
8. Thompson, 1966, p. 522.
9. Thomis, 1972, pp. 44–45.
10. Thompson, 1966, p. 201.
11. Thompson, 1966, p. 526.
12. Liversidge, 1972, p. 31.
13. Hills, 1973, pp. 62–63. See also Sale, 1995, p. 70n, for a longer list of "industrial actions."
14. Thompson, 1966, p. 553.
15. Ibid.
16. Thomis, 1972, pp. 17, 19, 20.
17. Liversidge, 1972, p. 42.
18. Standen, 1996, pp. 566–567.
19. Thompson, 1966, p. 192.
20. In the concluding lyric of the preface to his prophetic poem, *Milton*.
21. Thompson, 1966, pp. 192, 193.
22. Cardwell, 1972, p. 194.

23. Walter Licht, personal communication, July 17, 2001.

24. G. N. von Tunzelman, 1993, p. 269.

25. Eder, Richard, "He Gave Names to Clouds and Renown to Himself," *New York Times*, July 31, 2001, p. E6.

26. Ellul, 1964, pp. 229–318.

27. In Thomis, 1972, p. 50.

28. Shelley, 2000, p. 67. Reprint of the third edition, 1831.

## Chapter 2. Davy versus Stephenson

1. Whitaker, 1928, p. 74.

2. Quoted in Pohs, 1995, p. 268.

3. Hardwick and O'Shea, 1916, p. 551.

4. For a fascinating rundown on Coleridge's and Percy Bysshe Shelley's initial enchantment with Davy's scientific and philosophical ideas, and later, their diverging reactions to them, see Kipperman, 1998.

5. "Discourse, Introductory to a Course of Lectures on Chemistry Delivered in the Theatre of the Royal Institution," January 21, 1802 (London: Royal Institution, 1802), p. 22. Quoted in Kipperman, 1998, unpaged, obtained online.

6. In Knight, 1971, p. 600.

7. Carrier, 1965, p. 125.

8. In Carrier, 1965, p. 131.

9. Rolt, 1962, p. 6.

10. Rolt, 1962, p. 54.

11. Hobsbawm, 1996, p. 44.

12. Pohs, 1995, p. 304.

13. *Philosophical Magazine*, March 1817 (quoted in Smiles, 1859, pp. 109–110.)

14. Rolt, 1962, p. 30.

15. Rolt, 1962, p. 31.

16. This and the following three quotations are all from Rolt, 1962, pp. 32–33.

17. Knight, personal communication, October 15, 2001.

18. Both quotations from Smiles, 1859, p. 122.

19. Smiles, 1859, pp. 108–109.

20. Smiles, 1859, p. 124n.

21. David Knight, personal communication, October 26, 2001.

22. Hardwick and O'Shea, 1916, p. 597.

23. Smiles, 1859, p. 125.

24. Paul, 1924, p. 13.

25. *New York Times*, August 20, 2001, p. A8.

26. *New York Times*, September 25, 2001, p. A14.

27. Rolt, 1962, p. 11.

28. Rolt, 1962, pp. 14–15.

29. Both quotations from Rolt, 1962, p. 296.

30. D. M. Knight, personal communication, October 31, 2001.

31. In Carrier, 1965, p. 105.
32. Rolt, 1962, p. 321.
33. Pohs, 1995, p. 307.
34. Knight, 1971, p. 603.
35. Knight, 2000, p. 608.
36. Hartley, 1966, p. 121.

## Chapter 3. Morse versus Jackson and Henry

1. Mabee, 1943, p. 311.
2. Kendall, 1852, p. 3.
3. The following quotes are taken from Kendall, 1852. Granted, Kendall was not exactly a disinterested observer. Formerly postmaster of the United States, he had become an important part of the Morse team in 1845; as their business agent, he was vested with full powers to manage the company that had been formed by then. Still, the quotes appear to be genuine.
4. Kendall, 1852, p. 52.
5. Ibid., p. 54.
6. Ibid., p. 56.
7. Ibid., p. 58.
8. Ibid., p. 60.
9. All the *Post* quotes are from Kendall, 1852, p. 61.
10. Kendall, 1852, p. 14.
11. Mabee, 1943, p. 198.
12. David Hockfelder, personal communication, February 6, 2002.
13. Coulson, 1950, p. 215.
14. Moyer, 1997, p. 242.
15. Ibid., p. 333.
16. Ibid., p. 246.
17. Prime, 1875, p. 494.
18. Ibid., p. 494.
19. In a letter dated December 7, 1853. Vail, Colo., 1914, p. 23.
20. In Coulson, 1950, p. 227.
21. Moyer, 1997, p. 272.
22. Mabee, 1943, p. 311.
23. Smithsonian Institution, 1858, p. 89.
24. Coulson, 1950, p. 232.
25. Smithsonian Institution, 1858, p. 95.
26. Anonymous, 1872, p. 10.
27. Cooper, 1860, pp. 161–162. Prime's reference to the quote (p. 231) cites page 140, apparently referring to the original, 1849 edition.
28. Anonymous, 1872, p. 14.
29. For example, Mabee, 1843, p. 204.
30. Moyer, 1997, pp. 250–251.

## Chapter 4. Edison versus Westinghouse

1. Birdsall, 1979, p. 215.

2. Officially, George Westinghouse, Jr., though the Jr. is rarely used.

3. Writing admiringly of one of the railroad's engineers, a Mr. E. L. Northrop, the young newspaperman declared him to be "the most steady driver that we have ever rode behind [and we consider ourselves some judge, haveing been Railway riding for over two years constantly]."

4. Conot, 1979, p. 462, credits Rosanoff, M. A., "Edison in His Laboratory," *Harper's,* September, 1932.

5. Spelled Lucian in Prout, 1926.

6. Conot, 1979, p. 253.

7. Brown, July 14, 1889.

8. Anonymous, July 13, 1889.

9. Westinghouse, December 18, 1888.

10. Today, some high-voltage lines carry 100,000 volts or even more. Edison must be chuckling somewhere, for the magnetic fields generated by these lines have been accused of causing an increased rate of leukemia in children living under or near these lines. Later studies, however, have not borne out the early research.

11. Brown, July 14, 1889.

12. Anonymous, August 3, 1889.

13. Conot, 1979, p. 280.

14. Josephson, 1959, p. 360.

15. Tate, 1938, p. 278.

## Chapter 5. Ford versus Selden and ALAM

1. Lacey, 1986, p. 102.

2. This and the following quotes are from Ford, 1923, pp. 22–25.

3. Also referred to as Edison Lighting Co., Detroit Electric Co., Detroit Illuminating Co., and Detroit Edison Co.

4. Ford, 1923, p. 235.

5. Lacey, 1986, says that in the first few years of his employment, "Henry took over some spare space at the power station as a private workshop," and that he did much of his early experimenting there. Then, when things began to move along, he set up another workshop in back of his home (p. 41).

6. Temporary National Economic Committee, Monograph No. 31, "Patents and Free Enterprise," Washington, D.C.: Government Printing Office, 1941, pp. 115–119. Portion reprinted in Rae, 1969, pp. 96–98.

7. Greenleaf, 1961, p. 38.

8. Further details in Lacey, 1986, p. 61.

9. Lacey, 1986, p. 95.

10. Ford, 1923. p. 140.

11. Ibid., p. 140.

12. Ibid., p. 147.

13. Ibid., p. 184.
14. Ibid., p. 45.
15. Temporary National Economic Committee, op. cit., p. 96.
16. Bob Arnebeck, automobile historian, personal communication, May 30, 2002.
17. Nevins, 1954, p. 421. Additional details on the case can be found on pp. 416 and 418.
18. Lacey, 1986, p. 100.
19. Ibid., p. 101, cites *Detroit Journal*, Feb. 26, 1910.
20. Greenleaf, 1961, p. 118.
21. Ibid., p. 192.
22. Ibid., p. 169.
23. Ibid., p. 139.
24. Ibid., p. 147.
25. For more detail on the judicial proceedings, Chapter 7, "The Tournament of Motors," in Greenleaf, 1961.
26. Jardim, 1970, p. 92.
27. Greenleaf, in Rae, 1969, p. 137.
28. Temporary National Economic Committee, in Rae, 1969, p. 100.
29. From the Ford Archives of the Edison Institute, Dearborn, Mich. Cited in Lacey, 1986, p. 129.
30. For an interesting bit of history on the connection between automobile manufacturing and the creation of the middle class, see Redburn, 2000, p. 3.

## Chapter 6. Wright Brothers versus Curtiss, Chanute, Ader, Whitehead, and Others

1. Mouillard, 1892 (1881).
2. Chanute, 1894.
3. Langley, 1891.
4. Freedman, 1991, p. 29, says the Wrights observed buzzards. Howard, 1987, p. 33, says pigeons. Which one is right? It doesn't matter; the principle is the same.
5. Orville Wright, 1953, pp. 1168, 1169.
6. Scott, 1999, p. 101.
7. Ibid., p. 102.
8. In 1907, he was named the fastest man alive when he drove his motorcycle, powered by a V-8 engine, at the then-unbelievable speed of 136 mph at Ormond Beach, Florida.
9. Alberto Santos-Dumont, a Brazilian, had already done this in Europe two years earlier. He had learned of the Wrights' work from Chanute and flew a plane built by the Voisins.
10. Carpenter, 1992, p. 180.
11. "Glenn H. Curtiss: Founder of the American Aviation Industry." www.glennhcurtiss.com/id48.htm, unpaged (accessed April 2, 2002).
12. Loening, 1935, p. 48.

13. Scott, 1999, p. 161, 162.
14. New York *World,* December 12, 1909. In Howard, 1987, p. 339.
15. Howard, 1987, p. 331.
16. Ibid., p. 334.
17. Anonymous, 1914, p. 1.
18. Anonymous, 1914, p. 3.
19. Anonymous, 1914, p. 3.
20. Roseberry, 1972, p. 355.
21. Scott, 1999, p. 227.
22. Roseberry, 1972, p. 341.
23. Patent applied for by the AEA group November 1908. Although the courts felt that the Wright patent had priority, the Patent Office apparently had no problem with issuing a patent to the AEA group for their ailerons. It was granted on December 5, 1911.
24. Ettington, 2001, p. 2.
25. Roseberry, 1972, p. 334.
26. Krystek, 2001, p. 5.
27. www.firedragon.com/*kap/SteamTopics/whitehead.html (accessed April 2, 2002).

## Chapter 7. Sarnoff versus Farnsworth

1. *New York Times*, March 13, 1971, p. 32.
2. Godfrey, 2001, p. 111.
3. Lyons, 1966, p. 205.
4. Bilby, 1986, p. 173.
5. Stashower, 2002, p. 80.
6. Schwartz, 2002, p. 22.
7. Abramson, personal communication, July 11, 2002.
8. Stashower, 2002, p. 164.
9. Abramson, 1987, p. 199.
10. Ibid., p. 145.
11. Farnsworth, 1989, p. 145.
12. Schatzkin, Chronicles, Part 7.
13. Godfrey, personal communication, June 10, 2002.
14. Godfrey, 2001, pp. 57, 58.
15. Farnsworth, 1989, p. 158.
16. Abramson, 1987, p. 175.
17. Stashower, 2002, p. 255.
18. Trammell, 1959, p. 121y.
19. Lyons, 1966, pp. 57, 58.
20. Barnouw, 1990, pp. 17, 18. (Earlier editions, 1975, 1982.)
21. As for example in Lyons, 1966, pp. 100, 101.
22. Lyons, 1966, pp. 206, 208–210.
23. Dreher, 1977, p. 143.

24. Bilby, 1986, p. 128.
25. Anonymous, March 13, 1971, p. 32.
26. Anonymous, December 13, 1971, pp. 1, 43.
27. Anonymous, *New York Times,* December 24, 1971, p. 22.
28. Schwartz, 2002, p. 297.
29. Ibid., pp. 296, 297.
30. Schatzkin, online, "Who Invented Television?"
31. Godfrey, personal communication, June 10, 2002.
32. Bilby, 1986, p. 227.
33. Abramson, personal communication, July 11, 2002.
34. See bibliography: Godfrey, 2001; Stashower, 2002; Schwartz, 2002.

## Chapter 8. Rickover versus Zumwalt (and Just about Everyone Else)

1. For an excellent account, see Rockwell, 1992, pp. 87–90.
2. Craven 2001, e.g., p. 186.
3. Use of the words *atomic* and *nuclear* is sometimes confusing. Initially *atomic* was widely used. But it was quickly realized that the reactions involved are actually nuclear reactions, and so that term came to be more widely used, as in *nuclear-powered* ships.
4. Became part of the U.S. Department of Energy in 1975.
5. Blair, 1954, p. 120.
6. Frank Duncan, personal communication, July 25, 2002.
7. Polmar and Allen, 1982, pp. 327, 656, 657.
8. Frank Duncan, personal communication, August 26, 2002.
9. Mylander, 1980, p. 86.
10. Craven, 2001, p. 185.
11. Blair, 1954, p. 38.
12. Ibid., p. 18.
13. At the time, Zumwalt was not among the detractors. He felt that Rickover did deserve to be promoted. Frank Duncan, personal communication, July 25, 2002.
14. For more on this, see Polmar, 1982, pp. 191–194.
15. Blair, 1954, p. 208.
16. Polmar and Allen, 1982, p. 193.
17. Anderson, 1959.
18. Polmar and Allen, 1982, p. 176.
19. Frank Duncan, personal communication, July 29, 2002.
20. Tyler, 1986, p. 69.
21. Zumwalt, 1976, p. 88.
22. Ibid., p. 95.
23. Ibid., p. 64.
24. Ibid., p. 85.
25. Ibid., p. 74.

26. Ibid., p. 119.
27. Ibid., pp. 124, 125.
28. Ibid., p. 74.
29. Ibid., p. 121.
30. Ibid., p. 121.
31. Ibid., p. 155.
32. Duncan, 1990, p. 49.
33. Ibid., p. 49.
34. See, for example, Zumwalt, 1976, pp. 152–163.
35. Zumwalt, 1976, p. 197.
36. Ibid., p. 182.
37. *New York Times,* July 9, 1986, p. 26.
38. Duncan, 1990, p. 51.
39. Ibid., p. 51.
40. Ibid., p. xiv.
41. Zumwalt, 1976, p. 158.

## Chapter 9. Venter versus Collins

1. The name was changed to the National Human Genome Research Institute in 1993.
2. A natural hormone produced in the kidneys; it stimulates production of red blood cells and has a number of important medical uses.
3. See, for example, Regaldo, 2000, p. 50.
4. Adams, et al., 1991, p. 1656.
5. Davies, 2001, p. 60.
6. Cook-Deegan, 1994, p. 311.
7. Anonymous, *Economist*, April 19, 1992.
8. See, for example, Goode, 2000, pp. F1, F6.
9. Thompson, 1993, p. 23.
10. Mark Guyer, personal communication, September 28, 2002.
11. Larry Thompson, personal communication, September 28, 2002.
12. Cherfas, 2002, p. 42.
13. Adams, 1995.
14. Wade, May 10, 1998, p. 1.
15. The characters consist of varying sequences of four letters only. These are *a, c, t,* and *g,* referring to the four nucleotides, or bases that, in total, define both our form and function. The letters stand for adenine, cytosine, thymine, and guanine.
16. It later came out that one of these individuals was, not surprisingly, himself.
17. Marshall, 1999, p. 1908; and Mark Guyer, personal communication, September 28, 2002.
18. Davies, 2001, p. 150.
19. Ibid.
20. Belkin, 1998, p. 58, 59.

21. Marshall, 1999, p. 1907; and Davies, 2001, p. 153.
22. Larry Thompson, personal communication, September 28, 2002.
23. Mark Guyer, personal communication, September 28, 2002.
24. Anonymous, July 2000, pp. 94–97 (unpaged, obtained online).
25. Wade, May 17, 1998, p. 20.
26. Mark Guyer, personal communication, September 28, 2002.
27. Travis, 1998, p. 334.
28. Wade. 1998, p. 1.
29. Marshall, 1999, p. 1907.
30. Belkin, 1998, p. 31.
31. Preston, 2000, p. 66.
32. See, for example, Wade, June 22, 2000, p. A20.
33. Marshall, 2000, p. 2396.
34. Wade, May 18, 1999, p. F1.
35. Davies, 2001, p. 211.
36. Preston, 2000, p. 83.
37. Golden, 2000, p. 19.
38. Ibid., p. 19.
39. Anonymous, June 27, 2000, p. F8.
40. Ser Vaas, 2000, unpaged.
41. For some details of potential payoff see, for example, Chang, June 27, 2000, pp. F1, F7; and Altman, 2000, p. F6.
42. Anonymous, June 27, 2000, p. F8.
43. *Nature,* February 15, 2001, pp. 813–958.
44. Venter, February 16, 2001, pp. 1304–1351.
45. See, for example, Wade, June 27, 2000, pp. F1, F4.
46. Anonymous, June 27, 2000, p. F8.
47. Ibid.
48. Ibid.
49. Wade, May 2, 2001, p. A15.
50. Ibid.
51. Waterston, 2002, pp. 3712–3716.
52. Ibid., p. 3712.
53. Ibid., p. 3716.
54. Green, 2002, p. 4143.
55. Ibid.
56. Myers, 2002, p. 4145.
57. Green, 2002, p. 4143.
58. Ibid., pp. 4143–4144.
59. Ibid., p.4144.
60. Ibid.
61. Ibid.
62. Hensley, 2002, p. B6.
63. Pollack, 2002, p. 18.

64. See, e.g., Collins, 2003, unpaged.
65. Jasny and Kennedy, 2001, p. 1153.

## Chapter 10. Rifkin versus the Monsanto Company

1. www.monsanto.com/monsanto/layout/sci_tech/default.asp (undated; accessed September 27, 2002).
2. Rifkin, September 6, 1999, p. 12.
3. For example, T. Michael A. Wilson, chief executive of Horticulture Research International in England, states, "New technology has brought back the good old days for these Luddite propagandists." Letter to the Editor, *Wall Street Journal*, October 25, 1999, unpaged (obtained online). Also, " . . . apocalyptic vision of the Luddites . . ." in DeGregori, 2001, p. 22.
4. Stix, 1997, p. 28.
5. www.santafe.edu/*shalizi/notebooks/rifkin.html (undated; accessed September 27, 2002).
6. www.hypothesis.it/nobel/eng/wsw.htm (accessed September 4, 2002).
7. Stecklow, 1999, p 4.
8. Winston, 2002, p. 46.
9. Charles, 2001, p. 94.
10. Ibid., p. 95.
11. Rifkin, 1998, p. 99.
12. www.monsantodairy.com/farmer/index.html.
13. Pence, 2002, pp. 84–85, and Charles, 2001, p. 296.
14. Rifkin, 1998, p. 77.
15. Charles, 2001, p. 296.
16. See, for example, Travis, 2001, p. 68.
17. Harden, 1984, p. 122.
18. Rifkin, 1977, p. 110.
19. Hoyle, 1992, p. 1406.
20. van Biema, 1988, unpaged (obtained online).
21. Rifkin, 1977, pp. 9, 10. Rifkin dedicated the book to Huxley.
22. Stix, 1997, p. 32.
23. Rifkin, 1977, p. 142.
24. Rifkin, 1998, p. 116.
25. *The Meaning of It All: Thoughts of a Citizen-Scientist*. Reading, Mass.: Perseus Books, 1998, p. 120.
26. Rifkin, 1983, "Author's Note."
27. Harden, 1984, p. 189.
28. Rifkin, 1998, p. xiii.
29. Charles, 2001, p. 25.
30. Ibid., p. 204.
31. Rifkin, 1998, p. 79.
32. Ibid., p. 79.

33. More details in Lambrecht, 2001, pp. 34–37.

34. This and the following three quotes are from Freese and Giddings, 2001, p. 41.

35. For a brief report on EPA challenges of two companies' field tests, see Pollack, August 14, 2002, p. C5.

36. Winston, 2002, p. 118.

37. Freese and Giddings, 2001, p. 42.

38. Ibid.

39. Winston, 2002, p. 122; Freese and Giddings, 2001, p. 43.

40. Pollack, July 11, 2001, p. C8.

41. Freese and Giddings, 2001, p. 43.

42. For a friendly view of organic foods, see Pollan, May 13, 2001, pp. 30–37, 57–65.

43. Charles, 2001, p. 309.

44. Stecklow, 1999, unpaged (obtained online).

45. Stewart, 2002.

46. "UI President Urges Land-Grant Universities To Join Forces with Government and Industry To Battle Eco-Terrorism." University of Idaho press release, August 22, 2001.

47. Glickman, 2000, p. M5.

48. For a more extensive summary, see Eichenwald, 2001, pp. C1, C6.

49. Feder, 1999, p. C25.

50. Lambrecht, 2001, p. 244.

51. Barboza, 2000, p. C6.

52. Charles, 2001, p. 313.

53. Lambrecht, 2001, p. 36.

54. Pence, 2002, pp. 80, 81.

55. Charles, 2001, p. 261.

56. Monsanto Web site: "Who We Are Overview."

57. Condor, Bob. "Round 1 Defeats Rules for Gene-Altered Food." *Chicago Tribune,* November 10, 2002, p. 13. For a pro-labeling write-up on the referendum, see Forster, 2002. A short summary of biotech crop laws passed or in work in the United States for 2001 can be found in *Science News,* February 2, 2002, p. 77.

58. Coghlan, 2002.

59. Cauvin, 2002, p. A6.

60. "Force-Feeding the World," Item 12 in *News from The Campaign to Label Genetically Engineered Foods,* August 24, 2002 (www.thecampaign.org).

61. In *The Campaign Reporter Online,* September 4, 2002, Item 1 (www.thecampaign.org).

62. Ibid. For details on the Flavr Savr story, see Martineau, 2001, and Martineau, Spring 2001.

63. Winston, 2002, p. 245.

64. Chassy, personal communication, October 30, 2002.

65. More *GE News from The Campaign to Label Genetically Engineered Foods,* September 27, 2002, pp. 14–15 (www.thecampaign.org).

66. Freese and Giddings, 2001, p. 40.

67. Ibid., pp. 41.

68. More *GE News from The Campaign to Label Genetically Engineered Foods,* September 27, 2002, Item #2 (www.thecampaign.org).

69. Original article: Gould, 1985, pp. 34–38. Reprinted in *An Urchin in the Storm,* New York: W.W. Norton & Co, 1987. Quote is on page 230.

70. Gould, 1987, pp. 234–235, and p. 238.

71. See, for example, Donlan, 1992, p. 10.

72. For example, Travis, 1998, p. 299.

73. Barboza, December 15, 1999, p. C1.

74. For example, Brown, 1999; Barboza, December 15, 1999; Enserink, 1999; Anonymous, September 15, 1999; Cray, January/February 2000; Kilman, 1999.

75. A reference to Rifkin's 1983 suit against the University of California.

76. Anonymous (editorial), September 15, 1999, p. A32. See also Barboza, December 15, 1999.

77. Donlan, 1999, p. 58.

78. Pence, 2002, p. 197. For further discussion of the "Accident Story," see pp. 43–45 and 169–171.

79. Rifkin, June 7, 1999, p. 13.

80. Barboza, 1999, p. 4.

81. Steel, 2001, p. 3541.

82. Sears, 2001, p. 11937.

83. Ewen and Pusztai, 1999.

84. Brown, 1999, p 1089; Enserink, 1999, p. 656; Anonymous, Oct. 16, 1999, p. 85.

85. Anonymous, *Economist,* October 16, 1999, p. 85.

86. Davis, 1984, p. 44.

87. Otchet, 1998, p. 74.

88. Press release, September 20, 2002. Re ephedra, see Ives, 2003.

89. Pickrell, 2002, p. 26.

90. Schlosser, 2002, p. 159.

91. Stolberg, 2002, p. 25.

92. Winter is director of the FoodSafe Program at the University of California at Davis. His comments are from a recent *Expert Report* by the Institute of Food Technologists, and are summarized in a statement issued on November 5, 2002.

93. Bailey, 2001, unpaged (obtained online).

94. Winston, 2002, p. 77. For a further discussion of risk, see Brody, 2002, p. F11.

95. Dixon, 1999, p. 547.

96. Ibid., p. 548.

97. Cowen, 1998, p. 14.

98. Dixon, 1999, p. 548.

99. DeGregori, 2001, p. 18.

100. Charles, 2001, p. 313.

101. Van Biema, 1988, p. W13.

102. Rifkin, 1998, p. 234.

103. Ibid., p. 36.

104. Bailey, 2001, p. 35.

## Epilogue

1. The hieroglyphic script of the Egyptians was one of the earliest forms of writing.

2. Plato, 2000, pp. 171–172.

3. For more on this, see, for example, Cardwell, 1972, pp. 30–36, and Forbes, 1968, pp. 20–21.

4. Nelkin, 1995 (citation in General Background), p. 447.

5. Ferguson, 1973, p. 815.

6. Nelkin, 1995, p. 456.

7. Ibid., p. 449.

8. Ellul preferred this term over *technology* because of his strong feelings that just about everything in modern life had been subverted.

9. Ellul, 1967, p 306.

10. Ibid., p. 418.

11. Wilford, 2002, p. F6.

12. Pence, 2002, p. 199.

# Bibliography

## General Background

Akst, Daniel. "In Genoa's Noise, a Trumpet for Capitalism." *New York Times*, August 5, 2001, p. BU4.

Birdsall, Derek, and Cipolla, Carlo M. *The Technology of Man: A Visual History*. London: Wildwood House, 1980 (1979).

David, Edward E., Jr. "On the Dimensions of the Technology Controversy." *Daedalus,* Winter 1980, pp. 169–177.

Day, Lance, and McNeil, Ian. *Biographical Dictionary of the History of Technology*. New York: Routledge, 1996.

Ellul, Jacques. *The Technological Society.* New York: Alfred A. Knopf, 1967 (1964; original French edition 1954).

Hounshell, David A. *From the American System to Mass Production 1800–1932: The Development of Manufacturing Technology in the United States.* Baltimore: Johns Hopkins Press, 1984.

Jasanoff, Sheila, et al., eds. *Handbook of Science and Technology Studies*. Thousand Oaks, Calif.: Sage Publications, 1995.

Martin, Brian, and Richards, Eveleen. "Scientific Knowledge, Controversy, and Public Decision Making." Chapter 22 in Jasanoff, 1995, pp. 506–526.

McNeil, Ian. *An Encylopedia of the History of Technology.* London, New York: Routledge, 1990, pp. 624–625 (aeronautics); pp. 710–717, (telegraph).

Nelkin, Dorothy. "Science Controversies. The Dynamics of Public Disputes in the United States." Chapter 19 in Jasanoff, 1995, pp. 444–456.

Shulman, Seth. "IP's Bleak House." *Technology Review,* March 2001, p. 39. (Current patent nightmare.)

Singer, Charles Joseph, et al., eds. *A History of Technology.* Oxford: Clarendon Press, 1958. Vol. IV, pp. 89–98 (miner's lamp); pp. 644–662 (telegraph). Vol. V, pp. 208–234 (electrical industry); pp. 391–413 (aeronautics); pp. 418–437 (powered road vehicles).

Susskind, Charles. *Understanding Technology.* Baltimore: Johns Hopkins University Press, 1973.

Wamsley, James S. *American Ingenuity: Henry Ford Museum and Greenfield Village.* New York: Harry N. Abrams, 1985.

Warshofsky, Fred. *The Patent Wars: The Battle to Own the World's Technology.* New York: John Wiley & Sons, 1994.

## Introduction

Binswanger, Hans Christoph. "The Challenge of Faust." *Science,* No. 5377, July 31, 1998, pp. 640–641.

Krauthammer, Charles. "Return of the Luddites." *Time,* December 13, 1999, Vol. 154, pp. 24, 37.

## Chapter 1. Ned Ludd versus the Technological Revolution

AFN, "Luddite Influence in Mary Shelley's *Frankenstein.*" Online: www.afn.org/*afn31396/shelley.html. pp. 1–4 (accessed June 27, 2001).

Akst, Daniel. "Ludd's Choosy Children." *Technology Review,* January/February 1999, pp. 81–83.

Ashton, T. S. *The Industrial Revolution.* New York: Oxford University Press, 1997 (1968).

Bailey, Robert, and Taylor, Jack. "Textiles: The Water Frame." Online article, two pages, at www.woodberry.org (accessed July 31, 2001).

Brontë, Charlotte. *Shirley*. New York: The Modern Library, 1997 (1849). A long, slow-moving novel that gives a good idea of what life was like in the time of the Luddites.

Cardwell, D. S. L. *Turning Points in Western Technology*. New York: Science History Publications, 1972.

Crews, Roy. *The Valiant and the Damned*. New York: E.P. Dutton & Co., 1976. A novel based in the time of the Luddites. More romance than history, but gives a feel for the times.

Englander, David, and Downing, Taylor. Extracts from "The Mystery of Luddism." *History Today*, Vol. 38, March 1988. Unpaged; obtained online at www.thehistorychannel.co.uk/classroom/alevel/luds.htm.

Hawke, Gary. "Reinterpretations of the Industrial Revolution." Chapter 3 in O'Brien, 1993, pp. 54–78.

Hills, Richard L. *Richard Arkwright and Cotton Spinning*. Hove, UK: Wayland Publishers Ltd. 1973.

Levy, Steven. "The Luddites Are Back." *Newsweek*, June 12, 1995, p. 55.

Liversidge, Douglas. *The Luddites: Machine-Breakers of the Early Nineteenth Century.* New York: Franklin Watts, 1972.

O'Brien, Patrick, and Quinault, Roland, eds. *The Industrial Revolution and British Society*. New York: Cambridge University Press, 1993.

Philips, David. "Crime, Law and Punishment in the Industrial Revolution." Chapter 7 in O'Brien, 1993, pp. 156–182.

Pynchon, Thomas. "Is It O.K. to Be a Luddite?" *New York Times Book Review,* October 28, 1984, pp. 1, 40, 41.

Randall, Adrian. *Before the Luddites: Custom, Community and Machinery in the English Woollen Industry*. New York: Cambridge University Press, 1991.

Rossney, Robert. "The New Old Luddites. What's So Funny About Staying Alive?" *Whole Earth Review,* Spring 1994, n82, p. 2.

Sale, Kirkpatrick. *Rebels Against the Future: The Luddites and Their War on the Industrial Revolution*. Reading, Mass.: Addison-Wesley Publishing Co., 1995.

"Setting Limits on Technology. Lessons From the Luddites." *Nation*, June 5, 1995, pp.785–788.

Shelley, Mary. *Frankenstein, or The Modern Prometheus*. New York: Xlibris Corporation, 2000 (reprint of third edition, 1831).

Standen, Edith A. "History of Textiles." *Encyclopedia Americana,* Vol. 26, 1996, pp. 566–573.

Stevenson, John. "Social Aspects of the Industrial Revolution." Chapter 10 in O'Brien, 1993, pp. 229–253.

Thomis, Malcolm I. *The Luddites: Machine-Breaking in Regency England*. New York: Shocken Books, 1972 (hardcover 1970).

Thompson, E. P. *The Making of the English Working Class*. New York: Vintage Books, 1966.

Turner, Ronald, and Goulden, Steven N., eds. *Great Engineers and Pioneers in Technology* (Vol. 1, "From Antiquity through the Industrial Revolution). New York: St. Martin's Press, 1981.

von Tunzelman, G. N. "Technological and Organizational Change in Industry During the Early Industrial Revolution." Chapter 11 in O'Brien, 1993, pp. 254–282.

Watson, Bruce. "For a While, the Luddites Had a Smashing Success." *Smithsonian,* April 1993, pp. 140–154.

## Chapter 2. Davy versus Stephenson

Agricola, Georgius. *De Re Metallica*. Translated from the Latin edition of 1556. New York: Dover Publications, 1950 (pp. 200–211, on mine ventilation).

Beckett, Derrick. *Stephenson's Britain*. Newton Abbot, UK: David & Charles, 1984, pp. 33–34.

Carrier, Elba O. *Humphry Davy and Chemical Discovery*. New York: Franklin Watts, 1965.

Dorman, C. C. *The Stephensons and Steam Railways*. East Sussex, UK: Wayland Publishers, 1975.

Fullmer, J. Z. "The Poetry of Sir Humphry Davy." *Chymia*. Philadelphia: University of Pennsylvania Press, Vol. 6, 1960.

———. *Sir Humphry Davy's Published Works* (bibliography, specifically the years 1815–1818). Cambridge, Mass.: Harvard University Press, 1969, pp. 75–85.

Hardwick, F. W., and O'Shea, L. T. "Notes on the History of the Safety-Lamp." *Transactions—Institution of Mining Engineers*, Vol. 51, 1916, pp. 548–607.

Hartley, Sir Harold. *Humphry Davy*. London: Nelson, 1966, especially pp. 109–126.

Hobsbawm, Eric. *The Age of Revolution*. 1789–1848. New York: Vintage Books, 1996 (Original British edition, 1962.) Especially Chapter 2, "The Industrial Revolution."

Kipperman, Mark. "Coleridge, Shelley, Davy, and Science's Millennium. (Samuel Taylor Coleridge, Percy Bysshe Shelley, Humphry Davy)." *Criticism,* Summer 1998, Vol. 40, No. 3, pp. 409(28) (obtained online, via InfoTrac).

Knight, D. M. "Davy, Humphry." Encyclopedia entry in *Dictionary of Scientific Biography*, Vol. 3, 1971, pp. 598–604.

———. *Humphry Davy: Science and Power*. Cambridge, Mass.: Blackwell Publishers, 1992.

———. "Higher Pantheism." *Zygon*, September 2000, Vol. 35, No. 3, pp. 603–612.

———. "Humphry Davy: Science and Social Mobility." *Endeavour*, Vol. 24, No. 4, Winter 2000, pp. 165–169.

Paul, J. W., Ilsley, L. C., and Gleim, E. J. *Flame Safety Lamps*. Department of the Interior, Bureau of Mines, Bulletin 227, 1924, pp. 1–15.

Pohs, Henry A. *The Miner's Flame Light Book: The Story of Man's Development of Underground Light*. Denver: Flame Publishing Company, 1995.

Rolt, L. T. C. *The Railway Revolution: George and Robert Stephenson*. New York: St. Martin's Press, 1962 (first edition, 1960).

Smiles, Samuel. *The Life of George Stephenson, Railway Engineer*. Chicago: Belford, Clarke & Co., "From the fourth London Edition," 1859.

Smith, Martin Cruz. *Rose*. New York: Random House, 1996. A work of fiction. Set in a mining area in the nineteenth century, it gives some idea of what a miner's life was like at that time. There's some discussion about miners' lamps and firedamp.

Thesing, William B. *Caverns of Night: Coal Mines in Art, Literature, and Film*. Columbia: University of South Carolina Press, 2000. Critical essays of portrayals of British and American coal mines in painting, fiction, poetry, song, and film.

Treneer, Anne. *The Mercurial Chemist. A Life of Sir Humphry Davy*. London: Methuen & Co., Ltd., 1963.

Whitaker, J. W. *Mine Lighting*. London: Methuen & Co., 1928, pp. 73–81.

Williams-Ellis, Amabel, and Willis, Euan Cooper. *Laughing Gas and Safety Lamp: The Story of Sir Humphry Davy.* New York, Abelard-Schuman, 1954.

## Chapter 3. Morse versus Henry

Beauchamp, K. G. *A History of Telegraphy: Its Technology and Application*. London: Institution of Electrical Engineers, 2001.

Coe, Lewis. *The Telegraph: A History of Morse's Invention and Its Predecessors in the United States*. Jefferson, N.C.: McFarland & Co., 1993.

Cooper, James Fenimore. *The Sea Lions; or, The Lost Sealers*. New York, W.A. Townsend and Company, 1860. (Original publication: New York: Stringer and Townsend, 1849.)

Coulson, Thomas. *Joseph Henry: His Life and Work*. Princeton, N.J.: Princeton University Press, 1950.

Fagen, M. D., ed. *A History of Engineering and Science in the Bell System: The Early Years (1875–1925)*. Murray Hill, N.J.: Bell Telephone Laboratories, 1975.

Fahie, J. J. *A History of Electric Telegraphy, to the Year 1837. Chiefly Compiled from Original Sources, and Hitherto Unpublished Documents*. London and New York: E. & F.N. Spon, 1881.

*History Getting Right on the Invention of the American Electro-Magnetic Telegraph.*

Washington, D.C., 1872. Republication in book form of an article that appeared in the Philadelphia Press, Jan. 24, 1872, and the Sunday Chronicle (Washington, D.C.), Feb. 4, 1872.

Jahns, Patricia. *Joseph Henry: Father of American Electronics*. Englewood Cliffs, N.J.: Prentice-Hall, 1970.

Kendall, Amos. *Morse's Patent: Full Exposure of Dr. Chas. T. Jackson's Pretensions to the Invention of the American Electro-Magnetic Telegraph*. Washington, D.C.: 1852.

Mabee, Carleton. *The American Leonardo: A Life of Samuel F. B. Morse*. New York: Alfred A. Knopf, 1943. (Republished 1969 by Octagon Books, New York.)

Marland, E. A. *Early Electrical Communication*. New York: Abelard-Schuman, 1964.

Moyer, Albert E. *Joseph Henry: The Rise of an American Scientist*. Washington, D.C.: Smithsonian Institution Press, 1997.

Prime, Samuel Irenaeus. *The Life of Samuel F.B. Morse, LL. D., Inventor of the Electro-Magnetic Recording Telegraph*. New York: D. Appleton and Company, 1875.

Riedman, Sarah R. *Trailblazer of American Science: The Life of Joseph Henry*. Chicago: Rand McNally & Co., 1961.

Routledge, Robert. *Discoveries and Inventions of the 19th Century*. New York: Crescent Books, 1989 (reprint of 1890 edition), pp. 444–463.

Smithsonian Institution. *Annual Report of the Board of Regents of the Smithsonian Institution, Showing the Operations, Expenditures, and Condition of the Institution for the Year 1857*. Washington, D.C.: William A. Harris, Printer, 1858.

Standage, Tom. *The Victorian Internet: The Remarkable Story of the Telegraph and the Nineteenth Century's On-line Pioneers*. New York: Walker & Co., 1998.

Taylor, William B. *An Historical Sketch of Henry's Contribution to the Electro-Magnetic Telegraph: With an Account of the Origin and Development of Prof. Morse's Invention*. (From the Smithsonian Report for 1878. Washington, D.C.: Government Printing Office, 1879.)

Thompson, Robert Luther. *Wiring a Continent: The History of the Telegraph Industry in the United States, 1832–1866*. Princeton, N.J.: Princeton University Press, 1947.

Vail, Alfred. *The American Electro Magnetic Telegraph: With the Reports of Congress, and a Description of All Telegraphs Known, Employing Electricity or Galvanism*. Philadelphia: Lea & Blanchard, 1845.

———. *Early History of the Electro-Magnetic Telegraph from Letters and Journals of Alfred Vail, Arranged by His Son, J. Cummings Vail*. New York: Hine Brothers, 1914.

## Chapter 4. Westinghouse versus Edison

Adair, Gene. *Thomas Alva Edison: Inventing the Electric Age*. New York: Oxford University Press, 1996.

Anonymous. "Electricide: Two Words Suggested for Execution by Electricity." NY: *Buffalo Courier*, August 3, 1889. TAED SM007086a (accessed March 2, 2002).

Anonymous. "Is Electricity Sure to Kill?" Newark *Advertiser*, July 13, 1889. TAED SM007087b (accessed Dec. 12, 2001).

Baldwin, Neil. *Edison: Inventing the Century*. New York: Hyperion, 1995.

Bazerman, Charles. *The Languages of Edison's Electric Light*. Cambridge, Mass.: MIT Press, 1999.

Brown, Harold P. "Death Currents: Why Contact with Electric-Light Wires Proves Fatal," *New York Journal,* July 14, 1889, TAED SM007088a (accessed January 8, 2002).

Cheney, Margaret. *Tesla: Man Out of Time*. New York: Dell Publishing, 1983 (copyright 1981).

Clarke, Ronald W. *Edison: The Man Who Made the Future*. New York: G. P. Putnam's Sons, 1977.

Conot, Robert. *A Streak of Luck: The Life and Legend of Thomas Alva Edison*. New York: Seaview Books, 1979.

Dunham, Montrew. *George Westinghouse, Young Inventor.* Indianapolis: Bobbs-Merrill Company, 1963.

Edison, Thomas A. *The Diary of Thomas A. Edison*. (Introduction by Kathleen L. McGuirk.) Old Greenwich, Conn.: The Chatham Press, undated (1970?). Written during July 1885.

———. *The Papers of Thomas A. Edison*, edited by Reese V. Jenkins and others. Baltimore: Johns Hopkins Press, Vol. 1, 1989; Vol. 4, 1999.

Garabedian, H. Gordon. *George Westinghouse: Fabulous Inventor*. New York: Dodd, Mead & Company, 1943.

Jehl, Francis. *Menlo Park Reminiscences*. Dearborn, Mich.: Edison Institute, Vol. 1, 1937.

Josephson, Matthew. *Edison: A Biography*. New York: McGraw-Hill Book Company, 1959.

Pretzer, William S. "The Languages of Edison's Electric Light." Review of Bazerman, 1999. *Technology and Culture,* January 2001, pp. 171–173.

Prout, Henry G. *A Life of George Westinghouse*. New York: Charles Scribner's Sons, 1926.

TAED: Thomas A. Edison Papers Digital Edition. 1996–. Piscataway, N.J.: Rutgers University. www.edison.rutgers.edu (accessed January 8, 2002).

Tate, Alfred O. *Edison's Open Door: The Life Story of Thomas A. Edison, a Great Individualist*. New York: E.P. Dutton & Co., 1938.

Westinghouse, George Jr. "The Special Danger of Electric Lighting by Alternating Currents," *New York World.* December 18, 1888. TAED SM007069a (accessed January 8, 2002).

## Chapter 5. Ford versus Selden

Bruckberger, R. L. "The Ford Revolution." Chapter 17 in Rae, 1969, pp. 156–162.

Burlingame, Roger. *Henry Ford*. New York: Alfred A. Knopf, 1954.

———. "A Man Apart." Chapter 14 in Rae, 1969, pp. 142–147.

Faulkner, William. *The Reivers: A Reminiscence*. New York: Random House, 1962. Faulkner's fictional take on the early days of the automobile in Yoknapatawpha County.

"'Ford of the Air' Soon to Be Ready." *New York Times,* October 7, 1925, p. 1.
Ford, Henry, with Samuel Crowther. *My Life and Work*. Garden City, N.Y.: Doubleday, Page & Co., 1923.
Galbraith, John Kenneth. "A Relentless and Avid Self-Advertiser." Chapter 16 in Rae, 1969, pp.151–155.
Greenleaf, William. *Monopoly on Wheels: Henry Ford and the Selden Automobile Patent.* Detroit: Wayne State University Press, 1961.
———. "Ford and the Patent System." Chapter 13 in Rae, 1969, pp. 135–141. (Abstracts from Greenleaf, 1961.)
Jardim, Anne. *The First Henry Ford: A Study in Personality and Business Leadership*. Cambridge, Mass., 1970.
Kaempffert, Walter. "The Mussolini of Highland Park." *New York Times Magazine*, Jan, 8, 1928, page 1.
Lacey, Robert. *Ford: The Men and the Machine*. Boston: Little, Brown and Co., 1986.
"Millions Depend on Patent Suit." *New York Times,* March 18, 1928, sec. 4, p. 14.
Nevins, Allan. *Ford: The Times, the Man, the Company*. New York: Scribner, 1954.
Nevins, Allan, and Hill, Frank E. "The Most Spectacular Career in American Industrial Society." Chapter 15 in Rae, 1969, pp. 148–150.
Olson, Sidney. *Young Henry Ford: A Picture History of the First Forty Years*. Detroit: Wayne State University Press, 1963.
Rae, John B., ed. *Henry Ford*. Englewood Cliffs, N.J., 1969.
Redburn, Tom. "A Revolution Built in Mr. Ford's Factory." *New York Times,* January 2, 2000, Section 3, p. 4.
Sward, Keith. *The Legend of Henry Ford*. New York: Rinehart & Co., 1948.

## Chapter 6. Wright Brothers versus Curtiss, Chanute, Ader, Whitehead, and Others

Beddow, Reid. "Flight Into History," *Washington Post,* June 28, 1987, page X1 (Review of Howard, 1987.)
Callander, Bruce D. "The Wrights and Their Rivals." *Air Force Magazine*, August 1994, p. 68. (Review of Howard, 1987.)
Carpenter, Jack. *Pendulum: The Story of America's Three Aviation Pioneers, Wilbur Wright, Orville Wright and Glenn Curtiss, the Henry Ford of Aviation.* Carlisle, Mass.: Arsdalen, Bosch & Co., 1992.
Champlin, Charles. "Tom Swiftian Tale of America." *Los Angeles Times,* July 5, 1986, p. 1. (Some history about Curtiss.)
Chanute, Octave. *Progress in Flying Machines*. New York: The American Engineer and Railroad Journal, 1894. Reprinted 1976 (Long Beach, Calif.: Lorenz & Herwig).
Cooke, David C. *Who Really Invented the Airplane?* New York: G.P. Putnam's Sons, 1964.
Ettington, Raymond C. "The Patent War. Copy of the Original Wright Brothers' Patent of 1906 and the Bell, Baldwin, McCurdy, Curtiss, Selfridge (Aerial Experiment

Association) Patent of 1911, with a Brief History of the Wright vs. Curtiss Patent Fight, 1909–1914." Prepared for and available from The Glenn H. Curtiss Museum, Hammondsport, NY, June 6, 2001.

"Fliers Must Pay Him, Says Wright." *New York Times*, February 27, 1914, pp. 1, 3.

Freedman, Russell. *The Wright Brothers: How They Invented the Airplane.* New York: Holiday House, 1991.

Garfield, Eugene. "Negative Science." In *Essays of an Information Scientist,* Vol. 3, 1977–1978, pp. 155–166.

Gordon, Arthur. *The American Heritage History of Flight.* New York: American Heritage Publishing Company, 1962.

Grant, R. G. *Flight: 100 Years of Aviation.* New York: DK Publishing, Inc., 2002. Beautifully illustrated.

Howard, Fred. *Wilbur and Orville: A Biography of the Wright Brothers.* New York: Alfred A. Knopf, 1987.

Huntington, Tom. "Pioneering Virtues." *American History*, February 2001, p. 2. (A bit of personal history about the Wrights.)

Kirk, Stephen. *First in Flight: The Wright Brothers in North Carolina.* Winston-Salem, N. C.: John F. Blair Publisher, 1995.

Krystek, Lee. "Gustave Whitehead: Did He Beat the Wright Brothers into the Sky?" Online: www.unmuseum.mus.pa.us/gustave.htm, copyright 2001. (An excellent, balanced account of the claims and counterclaims.) (Accessed April 2, 2002.)

Langley, Samuel Pierpont. *Experiments in Aerodynamics.* Washington, D.C.: Smithsonian Institution, 1891 (reprinted 1902).

Loening, Grover. *Our Wings Grow Faster.* Garden City, N.Y.: Doubleday, Doran & Co., 1935.

———. *Takeoff Into Greatness: How American Aviation Grew So Big So Fast.* New York: G.P. Putnam's Sons, 1968.

Mackenzie, Catherine D. "How the Airplane Made Its Public Bow." *New York Times Magazine*, March 8, 1928, pp. 10, 11. (Credits F. W. Baldwin, an associate of Curtiss, with causing America to "quit laughing at flying machines." Also some background on ailerons.)

Mouillard, Louis-Pierre, *L'Empire de l'Air . . .* (Extract: *The Empire of the Air: An Ornithological Essay on the Flight of Birds.*) Washington, D.C.: *Annual Report of the Smithsonian Institution*, 1892, pp. 397–463.

Newcomb, Simon. "The Outlook for the Flying Machine." *The Independent: A Weekly Magazine*, Oct. 22, 1903, pp. 2508–2512. Reprinted in *Essays of an Information Scientist*, Vol. 3, 1977–1978, pp. 167–172.

Randolph, Stella. *Lost Flights of Gustave Whitehead.* Washington, D.C.: Places, Inc., 1937.

Roseberry, C. R. *Glenn Curtiss: Pioneer of Flight.* Garden City, N.Y.: Doubleday & Company, 1972.

Scott, Phil, ed. *The Pioneers of Flight: A Documentary History.* Princeton, N.J.: Princeton University Press, 1999. (In the words of the pioneers.)

Shulman, Seth. *Unlocking the Sky: Glenn Hammond Curtiss and the Race to Invent the Airplane.* New York: HarperCollins, 2002. (Leans to the Wright.)

Smith, Jack. "The Brothers Got All the Credit, but What About All Those Others with the Wright Stuff?" *Los Angeles Times,* December 15, 1986, p. 1. (Review of Howard, 1987.)

Villard, Henry Serrano. *Contact!: The Story of the Early Birds. Man's First Decade of Flight from Kitty Hawk to World War I.* New York: Thomas Y. Crowell Company, 1968.

Walsh, John Evangelist, *One Day at Kitty Hawk: The Untold Story of the Wright Brothers.* New York, Thomas Y. Crowell, 1975. (Fictionalized history; purports to show that Wilbur was the real brains behind the Wrights' success.)

Whitson, William W. *The Fledgling: Born for Flight.* Putnam Valley, N.Y.: Cogent Publishing, 2002. (A novel that mixes fact and fiction.)

Wright, Wilbur and Orville. *Miracle at Kitty Hawk: The Letters of Wilbur and Orville Wright.* New York: Farrar, Strauss and Young, 1951.

———. *The Papers of Wilbur and Orville Wright, including the Chanute-Wright Letters and other papers of Octave Chanute.* New York: McGraw-Hill, 1953.

Zahm, Albert Francis. *Early Powerplane Fathers: Henson, Goupil, Ader, and Whitehead.* Notre Dame, Indiana: University Press, 1945.

## Chapter 7. Sarnoff versus Farnsworth

Abramson, Albert. *The History of Television, 1880 to 1941.* Jefferson, N.C.: McFarland & Company, Inc., 1987.

———. *Zworykin, Pioneer of Television.* Champaign: University of Illinois Press, 1995.

Barnouw, Erik. *A History of Broadcasting in the United States.* Vol. 2, *The Golden Web, 1933 to 1953.* New York: Oxford University Press, 1968.

———. *Tube of Plenty: The Evolution of American Television.* New York: Oxford University Press, 1990. (Second revised edition. Original edition, 1975.)

Bilby, Kenneth. *The General: David Sarnoff and the Rise of the Communications Industry.* New York: Harper & Row, 1986.

Brinkley, Joel. "The Crime Behind Every TV." *New York Times Book Review*, June 9, 2002, p. 31. (Review of Schwartz, 2002, and Stashower, 2002.)

Burns, Russell. *John Logie Baird, Television Pioneer.* London: Institution of Electrical Engineers, 2000.

"David Sarnoff of RCA Is Dead; Visionary Broadcast Pioneer." *New York Times,* Dec. 13, 1971, pp. 1, 43.

DeMartino, Nick. "Restoring Philo's Place in History: Challenges to the Corporate Myth." Online at www.farnovision.com/chronicles/tfc-1977.html. (Originally published in *Televisions*, 1977.) (Accessed May 1, 2002.)

Dreher, Carl. *Sarnoff: An American Success.* New York: Quadrangle/New York Times Book Company, 1977.

Dunlop, Orrin E., Jr. "What Is Radio's Destiny for the Coming Decade?" (Predicts commercial television by 1940.) *New York Times*, January 12, 1930, Section 4, p. 15.

Everson, George. *The Story of Television: The Life of Philo T. Farnsworth*. New York: W.W. Norton, 1949.

Farnsworth, Elma. *Distant Vision: Romance and Discovery of an Invisible Frontier*. Salt Lake City: Pemberly Kent Publishers, 1989.

Fisher, David E., and Fisher, Marshall Jon. *Tube: The Invention of Television*. Washington, D.C.: Counterpoint, 1996.

Garbarine, Rachelle. "Billion Dollar Expansion Plan for High-Tech Center" (The Sarnoff Corporation). *New York Times*, February 4, 2001, p. NJRE 9.

Gladwell, Malcolm. "The Televisionary." *New Yorker*, May 27, 2002, pp. 82–87. (Essay review of Schwartz, 2002, and Stashower, 2002.)

Godfrey, Donald G. *Philo T. Farnsworth: The Father of Television*. Salt Lake City: University of Utah Press, 2001.

Gold, Glen David. *Carter Beats the Devil*. New York: Hyperion, 2001. Another indication that Farnsworth is coming out of the closet. Carter, a magician, meets Farnsworth, who plays a minor role in this novel. Carter uses Farnsworth's just-invented television to beef up one of his illusions.

Gould, Jack. "Sarnoff: Mr. Do-It of Broadcasting." *New York Times*, December 13, 1971, p. 43.

Henderson, Harry. *Communications and Broadcasting*. New York: Facts on File, Inc., 1997. (Chapter 6, pp. 99–116.)

Lyons, Eugene. *David Sarnoff: A Biography*. New York: Harper & Row, 1966.

McLean, Russell W. *Restoring Baird's Image*. London: Institution of Electrical Engineers, 2000.

O'Brien, Richard B. "An Immigrant Boy Climbs to the Top." (Profile of David Sarnoff.) *New York Times*, Jan. 12, 1930, Section 4, p. 15.

Ozersky, Josh. "Outside the Box." (Review of Stashower, 2002.) *Washington Post*. May 5, 2002, p. BW9.

"Philo T. Farnsworth, a Pioneer in Design of Television, Is Dead." *New York Times*, March 13, 1971, p. 32.

Pinsker, Beth. "Reinventing the Inventor." (Update on the Sarnoff Corporation.) *New York Times*, June 2, 2002, pp. NJ 1, 8.

Pollak, Michael. "Inside the Soap Opera of Television's Early Days." *New York Times,* Jan. 18, 2001, p. G8.

Schatzkin, Paul. *The Farnsworth Chronicles*. Eleven-part series, online at www.farnovision.com/chronicles/ (1977, 2001.) (Accessed May 1, 2002.)

———. "Who Invented Television?" Online at www:farnovision.com/chronicles/tfc-who_invented_what.html, 1977 (accessed May 1, 2002).

Schwartz, Evan I. "Who Really Invented TV?" *Technology Review,* September/October 2000, pp. 96–106.

Stashower, Daniel. *The Boy Genius and the Mogul: The Untold Story of Television*. New York: Broadway Books, 2002.

Trammell, Niles. "Radio and Television Broadcasting," *Encyclopedia Americana,* Vol. 23, 1959, p. 121y.

Udelson, Joseph H. *The Great Television Race: A History of the American Television Industry, 1925–1941.* Tuscaloosa, AL: University of Alabama Press, 1982.

## Chapter 8. Rickover versus Zumwalt

Anderson, William R. *Nautilus 90 North.* Cleveland: World Publishing, 1959.

Blair, Clay, Jr. *The Atomic Submarine and Admiral Rickover.* New York: Henry Holt and Company, 1954.

Craven, John Piña. *The Silent War: The Cold War Battle Beneath the Sea.* New York: Simon & Schuster, 2001.

Dao, James. "New Technology Enlisted to Save Aircraft Carrier." *New York Times,* May 20, 2001, p. 30.

David, Heather M. *Admiral Rickover and the Nuclear Navy.* New York: G.P. Putnam's Sons. 1970.

Duffy, Michael. "They Broke the Mold." (Obituary of Rickover.) *Time,* July 21, 1986, p. 27.

Duncan, Francis. *Rickover and the Nuclear Navy: The Discipline of Technology.* Annapolis, Md.: Naval Institute Press, 1990.

Finney, John W. "Rickover, Father of Nuclear Navy, Dies at 86." *New York Times*, July 9, 1986, pp. 1, 19.

Grossman, Karl. "Russian Submarine Tragedy Raises Questions About Nuclear Power." *Knight-Ridder/Tribune News Service*. August 18, 2000 (obtained online; p. K7012, article # CJ64783354).

Harris, Brayton. *The Navy Times Book of Submarines: A Political, Social and Military History.* New York: Berkley Books, 2001 (1997). (Contains two chapters on nuclear submarines within a broader context.)

Hewlett, Richard G., and Duncan, Francis. *Nuclear Navy, 1946–1962*. Chicago: University of Chicago Press, 1974.

Mintz, Jim. "How the Engineers Are Sinking Nuclear Power." *Science '83*, pp. 78–81 (online; accessed February 12, 2001).

Mylander, Maureen. "Requiem for a Heavyweight." *Washingtonian*, January 1980, pp. 86–89, 130–138.

Polmar, Norman, and Allen, Thomas B. *Rickover: Controversy and Genius.* New York: Simon and Schuster, 1982.

"The Rickover Record." *New York Times*, July 9, 1986, p. 26 (editorial).

Rickover, Hyman George. *Eminent Americans: Namesakes of the Polaris Submarine Fleet*. Washington, D.C.: Superintendant of Documents, U.S. Government Printing Office, 1972.

Rockwell, Theodore. *The Rickover Effect: How One Man Made a Difference*. Annapolis, Md.: Naval Institute Press, 1992.

Rodengen, Jeffrey L. *Serving the Silent Service: The Legend of Electric Boat.* Ft. Lauderdale, Fl.: Write Stuff Syndicate, 1994.

Shapley, Deborah. *Promise and Power: The Life and Times of Robert McNamara.* Boston: Little, Brown and Company, 1993.

Sullivan, George. *Inside Nuclear Submarines.* New York: Dodd, Mead & Company, 1982.

Thomas, Evan. "Overrun Silent, Overrun Deep: Probing Submarine Contracts and an Admiral's Gifts." *Time,* December 24, 1984, p. 14.

Tyler, Patrick. *Running Critical: The Silent War, Rickover, and General Dynamics.* New York: Harper & Row, 1986.

Zumwalt, Elmo R., Jr. *On Watch: A Memoir.* New York: Quadrangle/New York Times Book Co., 1976.

## Chapter 9. Venter versus Collins

Adams, M. D., et al. "Complementary DNA Sequencing: Expressed Sequence Tags and Human Genome Project." *Science*, June 21, 1991, pp. 1651–1656.

———. "The Genome Directory." *Nature*, September 28, 1995, pp. 1–379 (supplement).

Altman, Lawrence K. "Genomic Chief Has High Hopes, and Great Fears, for Genetic Testing." *New York Times,* June 27, 2000, p. F6. (Interview with Francis Collins.)

Arledge, Elizabeth, and Cort, Julia. *Cracking the Code of Life: The Race to Decode Human DNA.* (Nova videotape.) Boston, WGBH Boston Video, 2001.

Begley, Sharon. "Showdown in the DNA Corral." *Newsweek,* February 26, 2001, p. 62.

Belkin, Lisa. "Splice Einstein and Sammy Glick. Add a Little Magellan." *New York Times Magazine,* August 23, 1998, pp. 26–31, 56–60.

Chang, Kenneth. "Incomplete, Project Is Already Paying Off." *New York Times,* June 27, 2000, pp. F1, F7.

Chang, Laura. *Scientists at Work: Profiles of Today's Groundbreaking Scientists From Science Times.* New York: McGraw-Hill, 2000. Chapter on J. Craig Venter, pp. 64–71.

Cherfas, Jeremy. *The Human Genome: A Beginner's Guide to the Chemical Code of Life.* New York: DK Publishing, 2002.

Coghlan, A., and Kleiner, L. "On Your Marks . . ." *New Scientist,* May 23, 1998, p. 4.

Collins, Francis. "The Genome Doctor." Interview in *Christianity Today,* Oct. 1, 2001, pp. 42–46.

———. "Genomic Research." FDCH Congressional Testimony, May 22, 2003 (online, Master File Premier).

Cook-Deegan, Robert. *The Gene Wars: Science, Politics, and the Human Genome.* New York: W.W. Norton & Company, 1994.

Davies, Kevin. *Cracking the Genome: Inside the Race to Unlock Human DNA.* New York: The Free Press, 2001.

Galas, David J. "Making Sense of the Sequence." *Science,* February 16, 2001, pp. 1257–1260.

Gillis, Justin. "Gene Mapper's Stock Tumbles; Celera Shareholders Selling on the News." *Washington Post,* June 28, 2000, p. E1.

Golden, F., and Lemonick, M. D. "The Race Is Over." *Time,* July 23, 2000, pp. 18–23.

Goode, Erica. "Most Ills Are a Matter of More Than One Gene." *New York Times,* June 27, 2000, pp. F1, F6.

Green, Phil. "Whole-Genome Disassembly." *Proceedings of the National Academy of Science,* April 2, 2002, pp. 4143–4144.

Hazeltine, William A. "The Case for Gene Patents." *Technology Review,* September/October 2000, p. 59.

Hensley, Scott. "Venter Leaves Celera as Science, Business Clash." *Wall Street Journal*, January 23, 2002, pp. B1, B6.

Hensley, Scott, and Regaldo, Antonio. "Scientists Publish Critique of Celera's Work. Rivals Charge Firm Recycled Public Data in Genome Map." *Wall Street Journal,* March 5, 2002, pp. A2, A6.

Howard Hughes Medical Institute. *The Genes We Share With Yeast, Flies, Worms, and Mice: New Clues to Human Health and Disease.* Chevy Chase, Md.: HHMI Office of Communications, 2001.

Jasny, Barbara R., and Kennedy, Donald. "The Human Genome" (editorial). *Science,* February 16, 2001, p. 1153.

Jegalian, Karin. "The Gene Factory." *Technology Review,* March/April 1999, pp. 64–68.

Lander, Eric. "Riding the DNA Railroad." Interview in *Technology Review,* July 2000, pp. 94–97.

Lemonick, Michael D. "The Genome Is Mapped. Now What?" *Time,* July 3, 2000, pp. 24–29.

Marshall, Eliot. "A High-Stakes Gamble on Genome Sequencing." *Science,* June 18, 1999, pp. 1906–1909.

———. "Sharing the Glory, Not the Credit." *Science,* February 16, 2001, pp. 1189–1193.

Marshall, Eliot, et al. "In the Crossfire: Collins on Genomes, Patents, and 'Rivalry.'" *Science,* March 31, 2000, pp. 2396–2398.

Myers, Eugene W. "On the Sequencing and Assembly of the Human Genome." *Proceedings of the National Academy of Science,* April 2, 2002, pp. 4145–4146.

*Nature,* February 15, 2001. Special section on the human genome, pp. 813–958. (Analysis: pp. 813–859; articles: pp. 860–941; letters, pp. 942–958.)

O'Keefe, Christine L. "How to Close Those Pesky Genome Gaps." *Science* magazine online, October 4, 2001, pp. 1–5. www.sciencemag.org/feature/plus/sfg/articles/10_05okeefe.shtml.

Olshevsky, George. *Understanding the Genome*. New York: Warner Books, 2002.

Osborne, Randall. "Celera Resolute Despite Lawsuits, Human Genome Project Turmoil." *BioWorld Week,* June 5, 2000, p. 1 (online).

Pennisi, Elizabeth. "The Human Genome." *Science,* February 16, 2001, pp. 1177–1180.

Perry, Patrick. "Craig Venter: At the Helm of the Genetic Revolution." Interview in *Saturday Evening Post,* January 2000, pp. 48–54.

Pollack, Andrew. "Genome Pioneer Will Start Center of His Own." *New York Times,* Aug. 15, 2002, p. 18.

Preston, Richard. "The Genome Warrior." *New Yorker,* June 12, 2000, pp. 66–83.

Regaldo, Antonio. "The Great Gene Grab." *Technology Review,* September/October 2000, pp. 48–55.

Ridley, Matt. *Genome: The Autobiography of a Species in 23 Chapters.* New York: HarperCollins, 2000 (British edition, 1999).

Roberts, Leslie. "Genome Patent Fight Erupts." *Science,* October 11, 1991, pp. 184–186.

———. "Controversial from the Start." *Science,* February 16, 2001, pp. 1182–1188.

Roos, David S. "Bioinformatics—Trying to Swim in a Sea of Data." *Science,* February 16, 2001, pp. 1260–1261.

Ser Vaas, Cory, and Perry, Patrick. "For Dr. Craig Venter, Discovery Can't Wait." *Saturday Evening Post,* July/August 2000, pp. 38–45+ (obtained online).

"Shuffling Off the Coil. James Watson Resigns from Genome Program." *Economist,* April 18, 1992, p. 89.

Shulman, Seth. "Toward Sharing the Genome." *Technology Review,* September/October 2000, pp. 60–63, 66–67.

Thompson, L. "Healy and Collins Strike a Deal." *Science*, January 1, 1993, pp. 22–23.

Travis, John. "Another Human Genome Project. A Private Company's Plan Shocks the Genetics Community," *Science News,* May 23, 1998, pp. 334–335.

Venter, J. C. "Shotgun Sequencing of the Human Genome." *Science,* June 5, 1998, pp. 1540–1542.

Venter, J. C., et al. "The Sequence of the Human Genome." *Science,* February 16, 2001, pp. 1304–1351.

Wade, Nicholas. "Scientist's Plan: Map All DNA Within 3 Years." *New York Times,* May 10, 1998, pp. 1, 20.

———. "International Gene Project Gets a Lift." *New York Times,* May 17, 1998, p. 20.

———. "The Genome's Combative Entrepreneur." *New York Times,*" May 18, 1999, pp. F1, F2.

———. "Talk of Collaboration on Decoding of the Genome." *New York Times,* November 14, 1999, p. 22.

———. "Rivals in the Race to Decode Human DNA Agree to Cooperate." *New York Times,* June 22, 2000, p. A20.

———. "Double Landmarks for Watson: Helix and Genome." *New York Times,* June 27, 2000, p. F5.

———. "Now, the Hard Part: Putting the Genome to Work." *New York Times,* June 27, 2000, pp. F1, F4.

———. "Genome Feud Heats Up as Academic Team Accuses Commercial Rival of Faulty Work." *New York Times,* May 2, 2001, p. A15.

———. "Mouse Genome Is New Battleground for Project Rivals." *New York Times,* May 7, 2002, p. F2.

———. "Fish Genes Aid Human Discoveries." *New York Times,* July 26, 2002, p. A19.

Waterston, Robert H., et al. "On the Sequencing of the Human Genome." *Proceedings of the National Academy of Science,* March 19, 2002, pp. 3712–3716.

Watson, James. "The Double Helix Revisited." *Time,* July 3, 2000, p. 30.
"White House Remarks on Decoding of Genome," *New York Times*, June 27, 2000, p. F8.

## Chapter 10. Rifkin versus Monsanto

Bailey, Ronald. "Rage Against the Machines," *Reason,* July 2001, pp. 26–35.
Barboza, David. "Biotech Companies Take On Critics of Gene-Altered Food. Redesigning Nature—the Battle for Public Opinion." *New York Times,* November 12, 1999, p. 1.
———. "Monsanto Sued Over Use of Biotechnology in Developing Seeds." *New York Times,* December 15, 1999, p. C1.
———. "Industry Moves to Defend Biotechnology." *New York Times,* April 4, 2000, p. C6.
Brody, Jane E. "Risks and Realities: In a World of Hazards, Worries Are Often Misplaced." *New York Times,* August 20, 2002, p. F11.
Burros, Marian. "Chefs Join Campaign Against Altered Fish." *New York Times,* September 18, 2002, pp. F1, F7.
Butler, Declan. "Publication of Human Genomes Sparks Fresh Sequence Debate." *Nature,* February 15, 2001, pp. 747–748.
Carson, Rachel. *Silent Spring*. New York: Fawcett Crest, 1962.
Cauvin, Henri E. "Zambian Leader Defends Ban on Genetically Altered Foods." *New York Times,* Sept. 4, 2002, p. 6.
Charles, Daniel. *Lords of the Harvest: Biotech, Big Money, and the Future of Food.* Cambridge, Mass.: Perseus Publishing, 2001.
Coghlan, Andy. "Weeds Do Well Out of Modified Crops." *New Scientist,* August 17, 2002, p. 11.
Condor, Bob. "Round 1 Defeats Rules for Gene-Altered Food." *Chicago Tribune,* November 10, 2002, p. 13.
Cowen, Robert C. "Genetic Engineering of Pest-Resistant Plants Faces Public Resistance." *Christian Science Monitor,* May 13, 1998, p. 14.
Cray, Charlie. "Monsanto Sued." *Multinational Monitor,* January/February 2000, pp. 6, 7.
Davis, Bernard. "Ban Experiments in Genetic Engineering." *U.S. News & World Report,* October 8, 1984, p. 44.
DeGregori, Thomas R. "Genetically Modified Nonsense." Environment Unit Working Paper, Institute of Economic Affairs, London, January 1, 2001, pp. 1–32, (www.iea.org.uk/files/36.pdf).
Dixon, Bernard. "The Paradoxes of Genetically Modified Foods: A Climate of Mistrust Is Obscuring the Many Different Facets of Genetic Modification." *British Medical Journal*, February 27, 1999, pp. 547–548.
Donlan, Thomas G. "The Great Chefs: Cookery or Flackery?" *Barron's,* August 3, 1992, p. 10.
———. "A Visit From a Mob." *Barron's*, December 20, 1999, p. 58.

Eichenwald, Kurt. "Biotechnology Food: From the Lab to a Debacle." *New York Times,* January 25, 2001, pp. C1, C6.

Enserink, Martin. "*The Lancet* Scolded Over Pusztai Paper." *Science,* October 22, 1999, p. 656.

"EU/UN: Commission Tables Plan for Ratification of Biosafety Protocol. (European Union and United Nations to Ratify Protocol on Biosafety to the Convention on Biological Diversity, 2000.) *European Report*, March 16, 2002, p. 515.

Ewen, Stanley W. B. and Pusztai, Arpad. "Effects of Diets Containing Genetically Modified Potatoes Expressing *Galanthus Nivalis* Lectin on Rat Small Intestine." *Lancet,* October 16, 1999, pp. 1353–1354.

Feder, Barnaby J. "Plotting Corporate Futures. Biotechnology Examines What Could Go Wrong." *New York Times,* June 24, 1999, pp. C1, C25.

Forster, Julie. "GM Foods: Why Fight Labeling?" *Business Week,* November 11, 2002, p. 44.

Frankel, Mark. "Monsanto Faces a Fierce Food Fight." *Business Week,* December 27, 1999, p. 66.

Freese, William, and Giddings, L. Val. "Symposium: Q: Should Consumers Be Concerned About Bio-Engineered Crops?" *Insight on the News,* August 6, 2001, pp. 40–43.

Gianessi, Leonard P. *Plant Biotechnology: Current and Potential Impact for Improving Pest Management in U.S. Agriculture. An Analysis of 40 Case Studies.* National Center for Food & Agricultural Policy, June 2002.

Glickman, Dan. "Commentary: End the Biotech Food Fight; Agriculture: Let All Components of the Genetically Modified Food Debate Be Heard." *Los Angeles Times,* April 2, 2000, p. M5.

Gould, Stephen J. "On the Origin of Specious Critics." *Discover,* January 1985, pp. 34–38. Reprinted as "Integrity and Mr. Rifkin," in *An Urchin in the Storm,* New York: W.W. Norton & Co, 1987, pp. 229–239.

Harden, Blaine. "Of Genes and Little Boys." *The Washingtonian,* June 1984, pp. 118–123, 183–189.

Hoyle, Russ. "Rifkin Resurgent," *Bio/Technology*, November 1992, pp. 1406–1407.

Ives, Nat. "After a Pitcher's Death, Marketers of Dietary Supplements Try to Dodge the Taint of Ephedra." *New York Times,* March 17, 2002, p. 69.

Kilman, Scott. "Monsanto Is Sued Over Genetically Altered Crops." *Wall Street Journal,* December 15, 1999, p. 3.

Kolata, Gina. "Epidemic That Wasn't." *New York Times,* August 29, 2002, pp. B1, B5. (Breast cancer "epidemic.")

Lambrecht, Bill. *Dinner at the New Gene Café: How Genetic Engineering is Changing the Way We Eat, How We Live, and the Global Politics of Food.* New York: Thomas Dunne Books, 2001.

Martineau, Belinda. *First Fruit: The Creation of the Flavr Savr(TM) Tomato and Birth of Biotech Food.* New York: McGraw Hill, 2001.

———. "Food Fight (Genetically Engineered Foods)." *The Sciences,* Spring 2001, pp. 24–26.

McDade, Lucinda A. "Biotechnology in Agriculture: Good or Bad?" *Bioscience,* June 2002, pp. 534–536. (Essay review of Charles, 2001.)

Nesmith, Jeff. "Backgrounder: Genetically Modified Food: As Biotechnology Spreads, Questions Grow, Too." *Atlanta Journal-Constitution,* February 28, 2002, p. A7.

Otchet, Amy. "Jeremy Rifkin: Fears of a Brave New World." *UNESCO Courier,* September 1998, pp. 47–50.

Pence, Gregory E. *Designer Food: Mutant Harvest or Breadbasket of the World.* Lanham, Md.: Rowman & Littlefield Publishers, 2002.

Pickrell, J. "More Than Skin Deep?" *Science News,* July 20, 2002, p. 36.

Pollack, Andrew. "Gene Research Finds New Use in Agriculture." *New York Times,* March 7, 2001, pp. C1, C2.

———. "U.S. Finds No Allergies to Altered Corn." *New York Times,* June 14, 2001, p. C8.

———. "No Altered Corn Found in Allergy Samples." *New York Times,* July 11, 2001, p. C8.

———. "Data on Genetically Modified Corn. Reports Say Threat to Monarch Butterflies is 'Negligible.'" *New York Times,* September 8, 2001, p. C2.

———. "Earlier Safety Reviews Proposed for Gene-Altered Crops." *New York Times,* August 2, 2002. p. C3.

———. "E.P.A. May Fine Two Companies Over Tests of Engineered Corn." *New York Times,* August 14, 2002, p. C5.

Pollan, Michael. "Naturally. How Organic Farming Became a Marketing Niche and a Multibillion-Dollar Industry." *New York Times* magazine, May 13, 2001, pp. 30–37, 57–65.

Rifkin, Jeremy. *Who Should Play God?: The Artificial Creation of Life and What It Means for the Future of the Human Race.* New York: Delacorte Press, 1977.

———. *Algeny: A New Word—A New World.* New York: Penguin Books, 1984 (1983, The Viking Press).

———. *The Green Lifestyle Handbook. 1001 Ways You Can Heal the Earth.* New York: Henry Holt & Co., 1990.

———. *Voting Green: Your Complete Environmental Guide to Making Political Choices in the 1990s.* New York: Doubleday, 1992.

———. "Beware These Risky Foods." *USA Today,* March 18, 1996, p 12.

———. *The Biotech Century: Harnessing the Gene and Remaking the World.* New York: Jeremy P. Tarcher/Putnam, 1998.

———. "Commentary; Genetic Blueprints Aren't Mere Utilities; Biotechnology: We Can't Let a Few Conglomerates Control the Codes of Life and Trade Them as Commercial Goods." *Los Angeles Times,* July 8, 1998, p. 7.

———. "Unknown Risks of Genetically Engineered Crops." *Boston Globe,* June 7, 1999, p. 13.

———. "The Perils of the Biotech Century." *New Statesman,* September 6, 1999, p. 12–13.

Schlosser, Steve. "Blemishes" (letter to the editor). *Science News,* September 7, 2002, p. 159.

Sears, Mark K., et al. "Impact of Bt Corn Pollen on Monarch Butterfly Populations: A Risk Assessment." *Proceedings of the National Academy of Science, USA,* October 9, 2001, pp. 11937–11942.

"Seeds of Trouble." *Wall Street Journal* (editorial), September 15, 1999; p. A32.

Shiva, Vandana. *Stolen Harvest: The Hijacking of the Global Food Supply*. Cambridge, Mass.: South End Press, 2000.

Spurgeon, David. "Call for Tighter Controls on Transgenic Foods." *Nature,* February 15, 2001, p. 749.

Stecklow, Steve. "Germination: How a U.S. Gadfly and a Green Activist Started a Food Fight—Antibiotech Effort Bloomed Despite Little Funding and Lack of Consensus—'Who Should Play God?'" *Wall Street Journal,* November 30, 1999, p. 1.

Steel, Michael. "A Ticking Bomb in Your Corn Flakes?" *National Journal,* November 10, 2001, p. 3541.

Steinbrecher, Ricarda A. "From Green to Gene Revolution: The Environmental Risks of Genetically Engineered Crops." *Ecologist,* November/December 1996, p. 277.

Stewart, John. "The Race to Boost Organic Farming Is Heading Up a Dead-End Street." *The Scotsman,* August 27, 2002. Unpaged, obtained online: www.business.scotsman.com/agriculture.cfm?id=950862002.

Stix, Gary. "Profile: Jeremy Rifkin. Dark Prophet of Biogenetics." *Scientific American,* Aug. 1997, pp. 28–29.

Stolberg, Sheryl Gay. "Warnings Sought for Popular Painkiller." *New York Times,* September 20, 2002, p. 25.

"Trade—Caution Needed (International trade meeting in Montreal on genetically modified organisms)." *Economist,* February 5, 2000, p. 69.

Travis, John. "Patenting the Minotaur?" *Science News*; May 9, 1998, p. 299.

———. "Gene Makes Tomatoes Tolerate Salt." *Science News,* August 4, 2001, p. 68.

van Biema, David. "Biotech Gadfly Buzzes Italy." *Washington Post,* January 17, 1988. (Obtained online; start page, W13.)

Verhovek, Sam Howe, with Yoon, Carol Kaesuk. "Fires Believed Set as Protest Against Genetic Engineering." *New York Times,* May 23, 2001, pp. 1, 20.

Winston, Mark L. *Travels in the Genetically Modified Zone.* Cambridge, Mass.: Harvard University Press, 2002.

## Epilogue

Ferguson, Eugene. "Unassuaged Alarms." (Essay review of Susskind's *Understanding Technology; see* General Background for Susskind listing.) *Science,* November 23, 1973, pp. 815, 816.

Forbes, R. J. *The Conquest of Nature: Technology and Its Consequences*. New York: Frederick A. Praeger, 1968.

Plato. *Symposium and Phaedrus*. New York: Alfred A. Knopf, 2000. (Probably written in the 370s BCE. This translation is by Tom Griffith.)

Snow, C. P. *The Two Cultures.* New York: Cambridge University Press, 1959.

Wilford, John Noble. "Don't Blame Columbus for All the Indians' Ills." *New York Times,* October 29, 2002, pp. F1, F6.

# INDEX